The Pashmina

THIS BOOK IS
DEDICATED TO
MY WIFE TRIPTI

Who gave me the inspiration and
considerable support for writing the book

The Pashmina

Dr. Sailen Kumar Chaudhuri

WOODHEAD PUBLISHING INDIA PVT LTD

New Delhi

Published by Woodhead Publishing India Pvt. Ltd.
Woodhead Publishing India Pvt. Ltd.,
303, Vardaan House, 7/28, Ansari Road,
Daryaganj, New Delhi - 110002, India
www.woodheadpublishingindia.com

First published 2020, Woodhead Publishing India Pvt. Ltd.
© Woodhead Publishing India Pvt. Ltd., 2020

Woodhead Publishing India Pvt. Ltd. ISBN: : 978-93-88320-21-4
Woodhead Publishing India Pvt. Ltd. e-ISBN: 978-93-88320-22-1

Typeset by Bhumi Graphics, New Delhi

Contents

Preface

I have always had a wish to do write about Pashmina – The most exclusive textile fibre on Earth. But till 2010, I was extremely busy with the functioning of the wool mark company and could not devote the required time for this job. Now, that I am retired and had a quality time to go deep into the subject matter and subsequently a project done by me for ITC, UN for Nepal government on the study of pashmina status in Nepal and India. This project gave me immense support and knowledge to venture into writing of such a book. The book is relatively simpler version with bit of introductory subjects covering all aspects of the pashmina fibres and its products.

This book gives the reader a picture covering the Pashmina Industry and trade. Though, it is not a complete picture, but I definitely feel that this book will provide the reader with adequate information about Pashmina Fibre.

In this process, I must mention that quite a lot of information I had to borrow from several writers and contributors to the development of this Pashmina Industry. I have taken the liberty of borrowing some of the technical and marketing information through internet and many diversified literatures and I am grateful to all of them for this.

I hope that this first book on pashmina will be a useful and interesting to study for all those who would like to know about Pashmina.

Dr. S K Chaudhuri
New Delhi
13 July 2019

Acknowledgement

I must acknowledge the contribution of the following persons in bringing out this book. The contribution was of the type personal discussion, E-mail correspondences, by sharing the books and literatures, etc.

- Mr. Vijay Kumar Dugar, Secretary General of NPIA, Nepal. Manufacturer Care and Crafts, Nepal.

- Mrs. Mridula Jain, Chairperson of Shawl Club and of Shingora Shawls, India.

- Timir Roy, Secretary Indian National Office, Textile Institute.

- My Sons – Sougat and Suhrid Chaudhuri for their moral support and encouragement.

- Sourabh Mandal (Student of BPT) for giving me continuous support in editing and typing and other computer related issues.

- Dr. Phunchok, Former Senior Representative from Ladakh Autonomous Hill Development Council, Leh.

1
Introduction

Pashmina is also known as cashmere is a luxury fibre derived from Cashmere goat *Capra hircus* and many other goats from different parts of the world that are producing pashmina/cashmere. Although, cashmere and wool are two proteinase fibres with analogues chemical and physical attributes which makes pashmina as the most expensive fibre in the world. The high price of the fibre attracts adulteration with lower price fibre like wool or some manmade fibres.

The name pashmina has come from the word "Pashm", this is an Iranian word for pashmina.

The name 'Cashmere' may have come from the name Kashmir, which was created by British raj for their convenience. Pashmina and cashmere both are same structurally, property wise, hand, feel and touch. The pashmina has a bad name among the customers around the world which needs to be corrected. As a matter of fact, the finest wool in the world comes from goats which are native to Kashmir and palmer mountains.

The pashmina fibre is a kind of protein fibre that comes from downy undercoat of Cashmere goats which refer to different breeds of goats. The undercoat helps to keep the goats warms in temperature that can drop as low as −40°F in winter. Pashmina fibre is used to make shawls and knit wears. Now of course it is used in blends and 100% pashmina is also used for knitting wears. Pashmina was always seen as a pride possession, a luxury item and as an asset to the owner.

The largest group of these fibres is known as speciality hair fibre or luxury fibre. Figure 1 shows the classification of luxury fibres based on their sources.

The word pashmina is no longer defined as the fibre itself, but pashmina has become a brand name internationally of the end products. The production volume and quality have gone up considerably, lots of modernisation and innovation in the development of pashmina took place and the positive outcomes are also vividly seen. The birth of this fibre for creating so nice shawls and knit wears first started probably from Kashmir in India but slowly have gone into the countries which were much bigger in potentiality than India alone. Now China (70%) being the largest pashmina fibre producer in the world, Mongolia (18%) is the second, keeping Afghanistan (7%) third and distant producer of 1% fibre each is Nepal and India (Fig. 2).

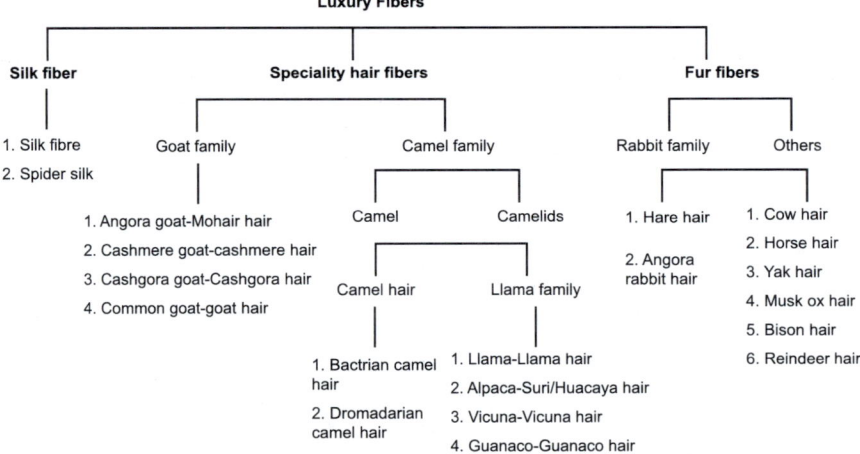

Figure 1: Luxury fibres

Figure 2: Pashmina fibre

The processing and the marketing of pashmina fibre and its product however are different in different countries. In one hand we see China as most advance processor of pashmina with full modernisation and on the other hand India and Nepal as we know are the classical producer of pashmina shawls. These are mostly done by hands only. Nature friendly shawls are defined as the product which use hand spun yarn and are woven in hand loom and naturally dyed with vegetable dyes. The ultimate product thus made is so nice in look and feel that anyone buying a product of this fibre forgets that it is made with most primitive methods and using most primitive tools. The production of pashmina fibre even now is too small in terms of overall demand in the world market. Pashmina is the fibre which is in demand and was in demand and will also be in demand. The cashmere fibre is also popularly known as **pashmina fibre**.

1.1   What is pashmina/Cashmere?

The pashmina fibre is the fine, down like undercoat of the Cashmere goat, which is biologically known as *Capra hircus* breed which is found mostly in the places where the temperature is extremely low and inhabitable. Pashmina goat growers are the biggest sufferer from cold, water shortage, lack of proper housing, especially India, Nepal, Afghanistan and some parts of China and the result is low production, but the qualities are not much affected. Production could be better if such issues are addressed rightly.

In countries like China, Mongolia and Afghanistan modernisation going on but the countries like India and Nepal are still on the basic and primitive way of pashmina production.

Figure 3, indicates that the Pashmina goats come in the micronaire range of 12 and 19 microns. It also indicates that within the goat family the pashmina producing goats have much finer fibre than others. Cashgora goats however is a misnomer as it is understood that it is a cross breed of Cashmere goat and Angora rabbits. The Chinese Cashmere goat is generally considered to produce much finer pashmina fibre which is of higher quality. The Mongolian cashmere is longer and coarser than Chinese pashmina fibre. Generally, the Afghanistan pashmina fibre is grouped together with the Iranian cashmere and is considered as the coarser than the Mongolian pashmina.

The definition used in United States of America

- Fine (dehaired) undercoat fibres produced by a Cashmere goat (Capra Hircus Laniger);
- Average diameter of such fibre does not exceed 19 microns; and
- Cannot contain more than 3% (by weight) of coarse cashmere fibres with average diameters that exceed 30 microns.

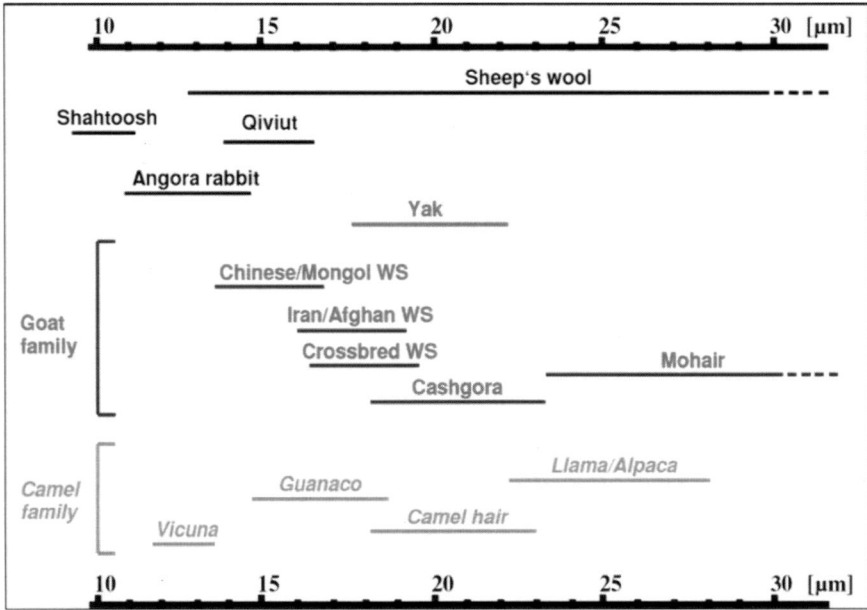

Figure 3: Fibre diameter of various natural animal fibres

Mislabelling

Due to lack of enough clarity in the definition of cashmere, there are lot of fraudulent and mislabelling going on in the market place.

Beside the fraudulent labelling of cashmere product there are customer deception as well with label specifying content and cashmere fibre blend. This misconduct in the business of cashmere end product ultimately affected the total business of pashmina and cashmere goods. These goods which were holding the higher place in the world market for long time and now with the various customer deceptions the word pashmina and cashmere have become totally confusing and unpopular. The cashmere industry in countries like

India, Nepal, Pakistan, the deception to the consumer was so rampant that the pashmina/cashmere industry becomes less effective.

According to the "US Federal Trade Commission" as with all other wool products, the fibre content of a shawl or other item marketed as pashmina must be accurately disclosed. For example, a blend of cashmere and silk might be labelled 50% cashmere, 50% silk or 70% cashmere, 30% silk, depending upon the actual cashmere and silk content. If the item contains only cashmere, it should be labelled 100% cashmere or all cashmere. The label cannot say 100% pashmina as pashmina is not a fibre recognised by the Wool Act or Regulation.

2.1 Early history of pashmina

Figure 4: Era of Mahabharata depicting pashmina shawls

Going back in the history, there is a mention in the Indian epic stories of **Mahabharata** in 3102 BCE that all the kings and their relatives using pashmina shawls abundantly (Fig. 4). In mid-eighteen century when European explorer and the employees of British India Company started carrying these shawls back home along with various stories like Napoleon gifted 17 shawls of this quality to his wife Josephine. Directly or indirectly these stories helped promoting the pashmina shawls in European countries. The shawls are very highly valued and seen as clear markers of status and asset of inheritance (Fig. 5).

Figure 5: Employees wives of British India Company with pashmina shawls

However, before the demand of cashmere shawls were made known by the Europeans, the Persian Empire distributed them as the Ropes of honours, the Mughal Empire establish the custom of endowing Kashmir shawls and to its alliance. Akbar, The Mughal emperor was probably the pioneer in initiating Pashmina Shawl production in Lahore and Agra (Fig. 6).

Figure 6: Persian Empire with pashmina shawls

The demand for cashmere shawl quickly raised and the demand could not be met. Britain and France were quickly up to the task and started imitating cashmere shawls. The main distinctive feature of this imitation cashmere shawl was the 'Buta' or cashmere pattern. With the change of fashion trend around 1817 Franco-Prussian War overall change in fashion and fall of price removed the royal image of these shawls.

2.2   Modern history of pashmina

The modern history of pashmina is not far, as it was only in 1998 the good times when cashmere shawl returned with full vigour. The price of cashmere shawls sold in late 1990s because of favourable economic environment or change in international fashion trend. It was marketed then in the name of 'Pashmina' as the luxury fibre and become among the most fashionable and desired item (Fig. 7).

Figure 7: Pashmina shawls in modern world

As such in the period of 1995–2002 there was an extraordinary boom in the overall shawl market. In India even the medical doctor left their profession and started making shawls as the shawl making was the most profitable business then. In one years' time shawl export became 10 times more with a fantastic amount of profit. The phenomenon changes the economy of Amritsar which is commercial capital of Punjab and famous for golden temple.

But unfortunately, the fashion and style have its own life cycle in marketplace. The business comes up to the peak and it crashes down, following this theory, the Cashmere shawl industry had its downturn sometimes around 2005. It was such a great fall that the industry found it difficult to deal with. They very badly needed the help for product and market diversification and large-scale product innovation.

The pashmina or the cashmere shawls were created perhaps to add a sensation of luxury to the product which worked well for quite some time till 2005. After the downturn, the cashmere product came back with some innovation on designing and fashion. But the turn was especially after the consumers were fed-up by the mislabelling and huge adulteration in the content of pashmina shawl.

Many of the shawls in the retail market are seen in the name of pashmina or cashmere shawls, pashmina did not even had a trace of cashmere fibre in the fibre content which was all totally synthetic but termed as pashmina shawls. The problem of pashmina is that if it is blended with fine merino wool, it is very difficult to identify the fibre content as there is no easy, proper and accurate way of testing. Those tests are available are very expensive and time consuming. The market for pashmina has gone down drastically because the consumers who can afford these products have a negative mind set for authenticity of the products. Their apprehension was justified also because of rampant misuse of the brand "Pashmina".

3

Pashmina – as the raw material

Pashmina was known in the western world as a shawl woven out of silk warp and a cashmere weft, a luxuries material. Pashmina is now known as a simple shawl with a soft touch maybe woven out of non-animal fibre. This is the confusion about the name pashmina the consumers have. This is a fallacy that pashmina which is originally a fibre and now that it has been taken in the market as a finished product especially in Nepal and India all over the globe. This book on pashmina though trying to be absolutely global on approach and international on dealing, quite often the examples are taken from various Indian studies.

Figure 8: Pashmina producing goats (Source of raw material)

A Cashmere goat is a type of goat that produces cashmere wool, the goat's fine, soft, downy, winter undercoat, in commercial quality and quantity. This undercoat grows as the day length shortens and is associated with an outer coat of coarse hair, which is present all the year and is called guard hair. Most common goat breeds, including dairy goats, grow this two-coated fleece. The down is produced by secondary follicles, the guard hair by the primary follicles (Fig. 8, Table 1).

Table 1: Pashmina producing goat

Name of the country	Pashmina/cashmere producing breed
Afghanistan	Asmari, Vatani
China	Chungwei, Liaoning
India	Changthangi, Chegu, Gaddi
Iran and Iraq	Kurdi, Morghose, Rani
Mongolia	Mongolian
Pakistan	Kaghani, Kel
Nepal	Northern Hill goat, Bhotia, Sinhal
Tibet	Tibetan
USSR (former)	Altai Mountain, Charihissar, Down, Don, Kirgis, Orenburg
Australia, New-Zealand, Scotland and USA	Feral Goats

Different types of goats from different countries

Australian Cashmere goat

Australian Cashmere goat is mainly seen in Northern and Western Australia. The production of these goats varies from herd to herd. The most productive herd are averaging 250 g of 15-micron cashmere fibre. There is a breed and fleece standards and active development continues with the University of Western Australia.

Changthangi Cashmere goat

The Changthangi and the pashmina goat are found in Ladakh and Baltistan (Kashmir region). They are raised for cashmere product. The breed is mostly white but black, grey animal also occurs. These goats have large twisting horns. This breed of goat produces cashmere fibre with average diameter of 12–13 microns and average fibre length between 55 and 650 mm. But its global production of this type of goat is almost 0.1% of global cashmere production.

Hexi

The Hexi Cashmere has a long history in desert and semi desert regions of Gansu province of China. About 60% of the goats are white. The Hexi Cashmere can be found in the Gansu, Qinghai and Ningxia. A typical adult do produces 184 g down having 15.7microns diameter.

Inner Mongolia Cashmere goats

This is a dual-purpose breed which adapts well to desert and semi desert pastures. These goats can be divided in the five stains Alasan (Alashanzuoqi), Arbus, Erlangshan, Hanshan and Wuzhumuqin. The average down yield is about 240 g with average diameter of 14.3 and 15.8 microns. The staple length would be 41 and 47 mm. Mongolian Goat population is about 2.3 million goats.

Liaoning Cashmere goat

Liaoning goat is mainly found in the Bayan Mountains in the Liaodong Peninsula. The breed was formally named the Liaoning Cashmere goat by the Chinese Ministry of Agriculture in 1984. By 1994, selected Liaoning does were producing 326 g of down at 15 μm diameter.

Licheng Daqing goat

The Licheng Daqing goat is a dual-purpose breed from the Shanxi Province, China. The down is usually brown, but the colour can vary. The average does down yield is 115 g at 14 µm diameter.

Luliang black goat

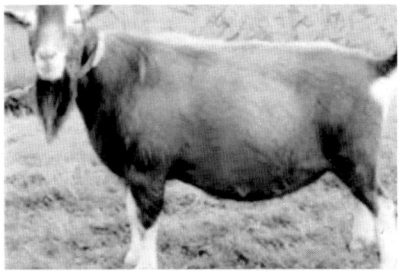

This dual-purpose goat is found in the Luliang area. It produces a small quantity of dark down.

Tibetan Plateau goat

In 1994, there were more than 7 million Tibetan Plateau and Valley goats in Tibetan Plateau regions of People's Republic of China. Five millions were in Tibet Autonomous Region, 1 million in Tibetan Autonomous Prefectures in Sichuan, half a million in Qinghai and about 100,000 in Gansu.

Totalling up to 13.6 million numbers of goats.

Wuzhumuqin

This Inner Mongolian strain is a new breed, recognised in 1994, and is distributed mainly in Xilingele Meng. The average production of a Wuzhumuqin adult does in 1994 was 285 g at 15.6µm diameter; the average down length was 46mm.

Zalaa Jinst White goat

The Zalaa Jinst White goat is the only entirely white breed of Cashmere goat in Mongolia recognised by the Mongolian Wool & Cashmere Association. The average cashmere production for males is 380 g; adult female is 290 g with fibres averaging 16.0–16.5 microns in diameter.

Zhongwei Cashmere goat

The Zhongwei goat originated in the semi desert and desert area around Zhongwei in Ningxia and Gansu Provinces in China, and are famous for their kid fur and cashmere production. The average fibre production for does is 216 g at 15 µm diameter.

Group of breeds

Table 2: Group of pashmina producing goats

Kirgis goat	Altai Mountain, Anatolian Black, Charhissar, Down, Don, Kazakh, Kirgis, Kurdi, Markhor, Orenburg, Uzbekh and Vatani
Mongolian goat	Chegtu, Chungwei, Jinning Liaoning, Mongolian, Wuan and Xingjiang
Kashmir goat	Changthangi, Chegu, Gaddi, Kaghani and Tibetan

3.1   Production

The total world production of the pashmina fibre is about 19,000metric tons annually.

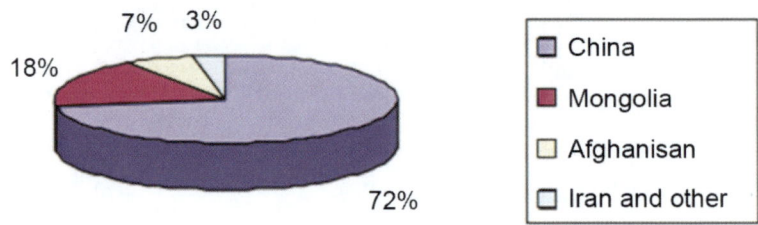

Figure 9: Global production of raw cashmere

As shown in the pie chart of the global production of raw cashmere, China is the largest producer of pashmina fibre in the world with 72% share (Fig.9). Whereas, Mongolia is at second position with 18% and Afghanistan is distant third with 7% share of the global pashmina fibre production. Surprisingly, the country where this fibre was born in the Himalayan range of India and Nepal, they are the tiny little partner of pashmina production with only 1% each. It has been described by the UN agencies (USAID). 'This colossal non-performance of India and Nepal has been studied and proper Actions have been taken appropriately by both the Governments' (Table 2).

World pashmina trade at a glance

Table 3: Production of pashmina of different countries

Sl. no	Parameters	Global	Nepal	India	China	Mongolia	Afghanistan
1	Popular name	NA	Tibetan Plateau goat	Changthangi and Chegu	Hexi, Licheng, Daqing, Liaoning, Zhongwei and Xinjiang	Alasan, Arbus, Erlangshan, Hanshan and Wuzhumuqin	Bahari cashmere, skin cashmere
2	% Share in the market		≤0.3%	≤1%	70%	20–28%	7%
3	Diameter	11 ≤ 19 μ	10–13 μ	9–20 μ	13–16 μ	15.5–17 μ	16–17 μ
4	Fibre length	21–40 m	35–40 mm	3–7 cm	4–5 cm	7–10 cm	19–25 mm
5	Fibre colour		Brown, grey	White, light grey, dark grey, brown	Predominantly white	60% Brown, 15% red, 10% light grey, 5% white	White cashmere, predominantly dark colour
6	Dehairing facility		Manual	Manual/ machine	Manual/ machine	Manual/ machine	Manual/ machine
7	Yarn quality		NA	Poor to very high end	Poor to very high end	Medium to high end	NA
8	Knitting quality		NA	Poor to very high end	Medium to high end	Medium to high end	NA
9	Shawl quality		Poor to upper medium end	Poor to high end	Poor to upper medium end	Poor to high end	NA
10	Value addition		Poor to upper medium end	Poor to high end	Poor to medium end	Poor to medium end	NA
11	Pashmina product diversification		Shawl stole, scarf	Shawls, scarf, stole	Scarves, stoles, shawls, knit wears	Scarves, stoles, shawls, knit wears	NA
12	Fibre in g/goat	Few grams to 750 g/goat	115–170 g	100–250 g	600 g	250–300 g	150 g
13	Annual production	11000–16000 tonnes	45–50 tonnes	45–50 tonnes	10000 tonnes	3300 tonnes	1000 MT
14	Export value of pashmina annually		27.8 million dollars (2013–14)	80 million dollars	1265 million dollars	180.7 million dollars	18 million USD in export of greasy cashmere

Contd...

Contd...

Sl. no	Parameters	Global	Nepal	India	China	Mongolia	Afghanistan
15	Countries it is exported		USA, Italy, Canada, UK, France, Japan, Germany	Major demand from Europe	India, Canada, Ireland, Italy, Slovenia, Spain, Switzerland, UK, US, Chile, France, Japan	Japan, China, USA, Switzerland	Raw pashmina mainly to Belgium, Iran, China, Dubai
16	R&D/ innovation activities	Nil	Low	High	High	High	NA
17	Future plan	High	High	High	High	High	NA

It is clearly seen from Table 3 above that the China is undoubtedly the leader in producing pashmina fibre.

3.1.1   Harvesting

The pashmina fibre as a raw material coming from the places which are generally high altitude with temperature going down to −40°C and it is not one day but normally few months in the year at a stretch. This terrible climatic condition becomes unbearable for the fibre growers, the goat herders and the goats. Until and unless we improve the lives of these herders with proper medical treatment facility, proper water supply for drinking and useful needs and adequate housing facilities, there is no way to improve the quality and the productivity of pashmina/cashmere fibres.

Figure 10: Harvesting

Cashmere goats are normally taken for combing operations. The process is like combing the tangled human hair using most primitive combs. In India, combing is the major method of harvesting using a special type of comb. After harvesting, pashmina fibre is dusted manually to remove some impurities like sand, dust, etc. which account about 10–20% of fibre weight (Fig. 10).

Typical Mongolian herders own about 100 goats. A man with 400 goats is considered relatively wealthy.

The cashmere producing goats are known to eat up by the root not leaving any vegetation while grazing, the top soil blows away, grassland turns to desert and dust storms creates atmospheric pollution. The problems of desertification and loss of pastureland compels the authority to have decimated the rate flocks and ordered more rational farming. As a result, cashmere buyers have turn to Mongolia for supplies pushing the price up in recent years.

The value of the pashmina fibre is determined by its diameter, staple length, colour and crimp. Currently the best prices are paid for white fibre for the staple length of more than 5 centimetres however, the best prices offered in Ladakh fall far short of real value and the small farmers remain hostage to the whims of the middle men until such time a competitive purchase is established, the situation is unlike to change for the better. Corporative could provide such competition and if effective would encourage small farmers to produce more pashmina. In addition to the staple length, colour and diameter other factors like the yield, cleanliness of the fibre will also matter for price realisation.

3.1.2   Structural and chemical properties

Pashmina fibre is an animal hair almost identical with fine merino wool and mohair fibre. It has predominance of ortho-cortical and meso-cortical cells. Meso-cortical cells have higher micro fibril packing density than wool fibre of the same diameter. This may be associated with low crimp pashmina fibre. It has bilateral structure and the percentage of ortho and para cortex is 50.4% and 49.6%, respectively. The number of scales per 100-micron range is 6.5–9 (Fig. 11).

Pashmina fibres are 10% weaker than fine Marino wool and 47% weaker than mohair fibre. The pashmina has a fibre fineness of 10–14 microns. The feel and handle as expected in a luxury fibre is very soft and warm. The amino acid composition of pashmina fibre is very similar to fine wool except cysteine, tyrosine (12% more than wool) and proline (9% lower than pashmina).

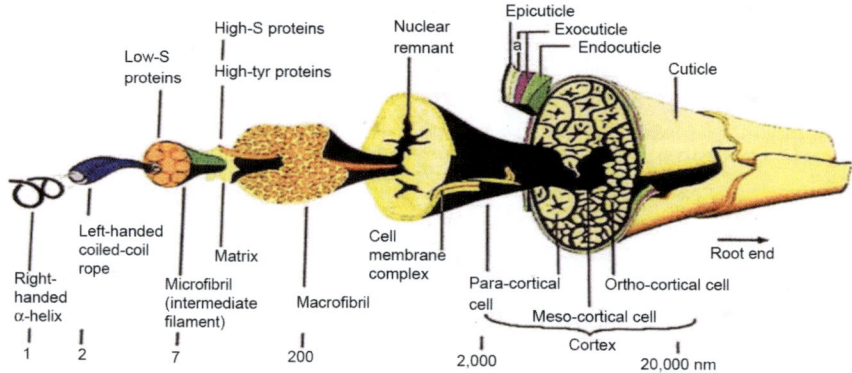

Figure 11: Structure of animal hair including wool and pashmina

Pashmina fibre is a protein fibre with polyamide polymer made up of 20 amino acid with alpha keratin, arranged in helical structure like that of wool (Fig. 12).

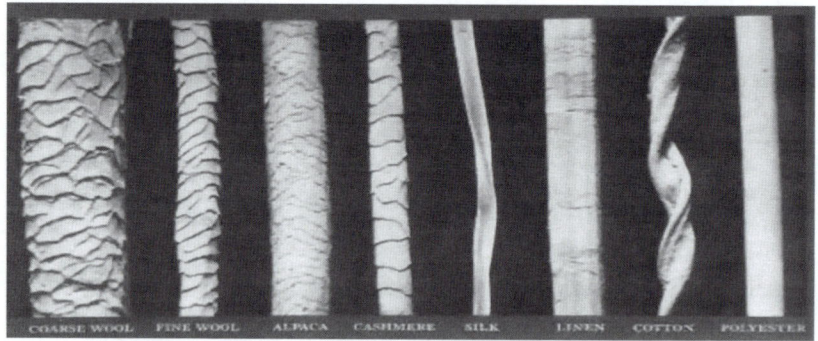

Figure 12: SEM of pashmina fibre along with some other fibres

It is observed by some of the researcher that the lipid composition is lesser than fine wool. It is also observed that the pashmina fibre has more of the polar amino acid like thionine, serine and tyrosine then fine wool making the cuticle more hydrophilic than fine wool. The chemical properties and physical properties are very similar to fine merino wool.

Recently, few good innovations are seen in the field of pashmina dyeing and finishing processes like nano-finishing, plasma treatment, enzyme treatment and aroma finishing (Table 4).

Table 4: Pashmina fibre characteristics from various breeds of goats from the world

Country	Breed	Fibre diameter	Staple length	Colour
Russia	Predonskays	18–19μ	8–12 cm	Grey
China	Xinjiang	15μ	4–5 cm	White
Mongolia	Koigurvansakhan	16–17μ	7–10 cm	Grey
Ladakh	Changra	10–15μ	3–7 cm	White
Nepal	Changthangi	10–13μ	3.5–4.0 cm	Brown, grey

3.1.3   Physical properties

The quality of pashmina fibre can be described best if its following physical properties are considered.

- Diameter
- Fibre length
- Colour
- Crimp
- Yield

Diameter (Pashmina Fibre)

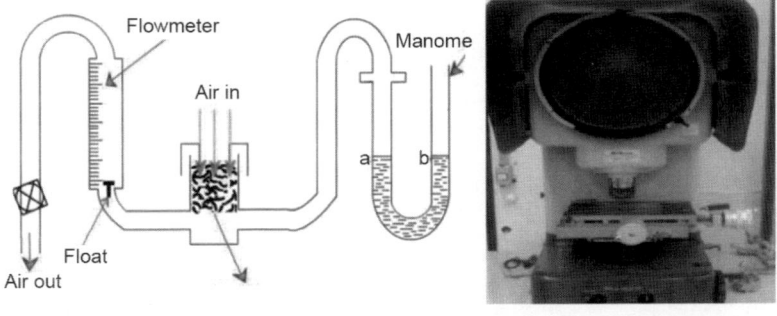

(a) Airflow method   (b) Projection microscopy

Figure 13: Diameter measurement by using (a) airflow method and (b) projection microscopy

Diameter of pashmina fibre is the most important characteristic on which quality and price are totally dependent. Lesser the diameter of the fibre, finer the fibre. The price of the fibre mainly depends on fineness of the fibre i.e. micronaire value.

The diameter of the fibre can be measured by using projection microscope as shown in Fig. 13 above.

Fibre length

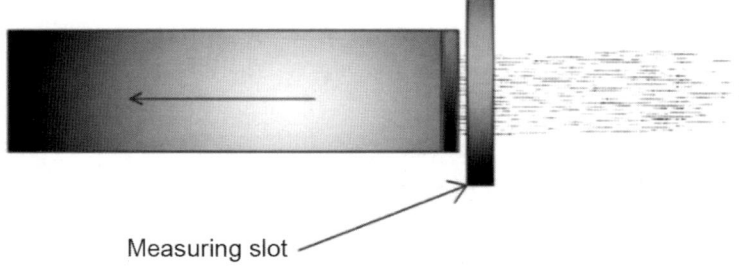

Measuring slot

Figure 14: Measurement of staple length of pashmina
fibre by using capacitive method

Fibre length which is really the staple length of the fibre is very important as far as the spinability is concerned. The shorter length will be a problem in spinning and then in the end products as the yarn will not be smooth and having lots of short fibre protruding making the yarn hairy (Fig. 14).

A hairy yarn is not acceptable. The longer fibre on the other hand will create problems in drafting zone breaking the fibre into unacceptable measurement to reach the price and quality equilibrium. It is rather advisable to follow the specification correctly.

Colour

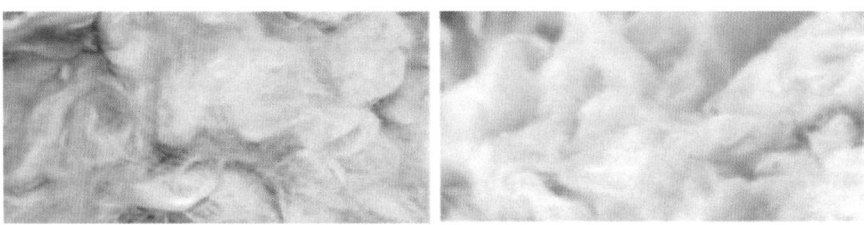

Figure 15: Natural colour difference in raw pashmina

The natural colour has also impact on the price of the pashmina fibre. Darker the colour of the Pashmina fibre, lesser will be the chances of making pastel shades which is a limitation and this limitation ultimately fetching lower price in the world market (Fig. 15).

Crimp

Crimp is an important quality criterion, as lesser the crimp, lesser the spin ability. This is because of the protein structure of the wool (alpha helix keratin), similar to the pashmina fibre. Crimps help in spinning better than straight fibres like Alpaca. The worsted industry demands to have 9 to 10 crimps per inch at least in the Australian wool they buy.

Yield

Yield is defined as the amount of clean and washed fibre we get out of a raw pashmina in terms of percentage. This has direct implication on the price of the fibre.

Table 5: Mechanical characteristics of changthangi pashmina

Characteristics	In air		In water	
	Fine Pashmina	Guard hair	Fine pashmina	Guard hair
Tex	0.319 ± 0.0072	3.004 ± 0.1106	0.318 ± 0.0072	2.864 ± 0.1024
Breaking strength	11.812 ± 0.1475	14.597 ± 0.2596	10.043 ± 0.1344	9.339 ± 0.2773
Breaking extension	34.58 ± 0.3774	37.54 ± 0.4408	55.93 ± 0.6019	55.38 ± 0.2773

Fibre fineness which is the function of the fibre diameter is the most important quality parameter in pashmina fibre because this property not only impact on the overall quality of the yarn and the end product but also help in determining the price of the fibre (Table 5).

Fibre length is an important characteristic that is directly linked with the spin ability. Therefore, it is important for spinner to specify the desired staple length. It is one of the important job of a spinner to specify the right specification otherwise under specification will have impact on the quality of the yarn negatively and over specification will have impact on the cost of the yarn directly.

There is a distinct relationship between the mechanical properties to the structure of fibre. The mechanical properties via cross linking between the filament and the matrix are largely responsible for observed mechanical properties of the fibre. This can be very well said as pashmina structure is very similar to wool so the properties which are discussed will be applicable to pashmina as well. The other characteristics colour, style and crimp are also equally important for determining prices. If the colour of hair is not good then it cannot be dyed in pastel shades and naturally the price of the fibre will be affected. The yield ratio determines the ratio of actual clean fibre counted after scouring and de-haring to that of raw pashmina.

The appeal and unrivalled status of pashmina fibre as the luxury fibre brings about three key factors:
- Fibre fineness (10–14 μ).
- Visual appearance and extreme softness.
- Keeping a big gap between supply and demand and its mysticism.

Chemically it is identical with wool and that is the reason the structural properties of wool are also considered for pashmina. Pashmina is 10% weaker than fine merino wool and about 40% weaker than mohair fibre.

3.2   Processing

The processes involved in the value addition of pashmina fibre (Fig. 16):

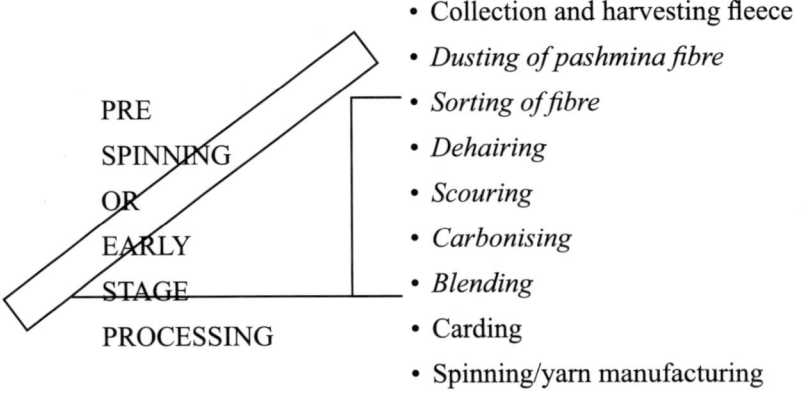

- Collection and harvesting fleece
- *Dusting of pashmina fibre*
- *Sorting of fibre*
- *Dehairing*
- *Scouring*
- *Carbonising*
- *Blending*
- Carding
- Spinning/yarn manufacturing

- Preparatory process for weaving
- Weaving and knitting
- Colouration and finishing
- Embroidery and design

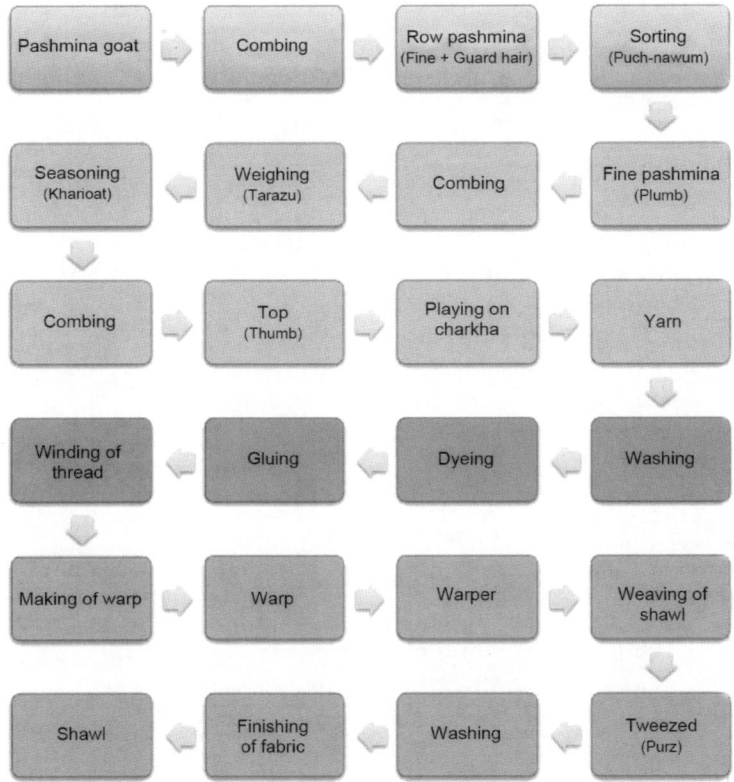

Figure 16: Traditional process sequence of shawl manufacturing in Kashmir

3.2.1   Collection and harvesting fleece

The collection of harvested pashmina fibre **is** a very technical Subject. The Pashmina fleece is collected either by combing or shearing. During every spring season each goat produces from 80 to 450 g of pashmina fibre. For manufacturing one pashmina shawl, it requires pashmina fibre from three goats (nearly 300–1000 g). The average fibre fineness is from 11.6 to 15.4μ and average length of the fibre which is staple length is 50 mm. The male sheep (Buck) yield more pashmina fleece than female sheep (Doe) (Fig. 17).

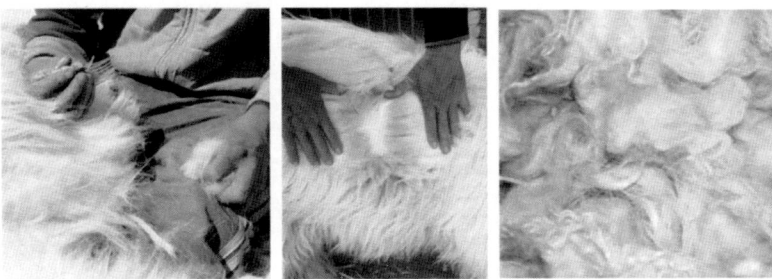

Figure 17: Collection and harvesting fleece

The countries that produce commercial quantity of pashmina are China with 70% of the world's output, outer Mongolia with 15–20%, Iran and Afghanistan at a combined 10–15%. There are some small cashmere producers in Central Asian countries i.e. kazakhstan, Kyrgyzstan, Tajikistan and Turkey, India, Australia and Pakistan but these quantities are small in comparison to three primary producers.

3.2.2   Dusting

The process of dusting is equally important, and it must be done before sorting (Fig. 18).

Figure 18: Dusting of pashmina fibre

Dusting process will take out all the attached impurities like dust, vegetable matters, dungs, etc. This process loses 20–80% of fibre weight and is done manually.

3.2.3   Sorting

The pashmina fleece from anywhere in the world goes for the process of sorting before dehairing. Sorting is also done after spinning of pashmina yarns. The fleece can be sorted out manually based on the fibre length, colour and quality. Sorting process can also be applied after pre spinning stage (Fig. 19).

Figure 19: Sorting of pashmina fibre

In China and Mongolia the first step to sort out low grade cashmere and if any synthetic contamination that is found, the stock is then baled and stored; the second sorting comes when we are ready to breakdown the fibre into colours like white, light grey, cream and brown. During the second sorting normally, it is again looked for any lower grade fibre or synthetic contamination that got through the first sorting.

3.2.4   Dehairing

Dehairing is a process by which outer coat of guard hair is separated from the under coat fine fibres (Fig. 20).

Figure 20: Dehairing by hands

Raw pashmina fibre is having 50–60% of guard hairs. The guard hairs are very thick with 50–100microns diameter, covering the fine down fibre of 10–13microns. The guard hairs should be removed completely before processing or improving spinability or developing best quality end products. The presence of more than 5% guard hairs will definitely affect the appearance, hand feel and quality of final product. The dehairing process was previously carried out in India, Nepal and Afghanistan not using any machine but by hands.

Figure 21: Dehairing of pashmina fibre by machine use

The process was time consuming and laborious. Since guard hair is much coarser than under coat fine fibre, proper elimination is essential for better processing. The modified cotton card is also used in dehairing machine that removes the guard hair from soft under coat. In market, 3% residual guard hair can be allowed and legally considered as pashmina fibre for export. 0.5% is acceptable for weaving of shawl and 0.2% or lower level for knitting. The left-over coarse hair/modulated and kempy fibres adversely affect the product quality and price of the pashmina fibre. The price of kempy fibres is about 30USD (Fig. 21).

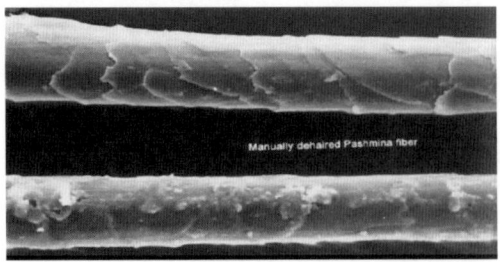

Figure 22: Manual dehaired pashmina fibre

A study was conducted sometime back in India to understand if there is any difference between – **manual dehair** and **machine dehair** fibre. It was observed that fibre diameter showed no significant difference whereas fibre length did show significant difference. The Difference is that the machine dehairing breaks more fibre and produces a considerable percentage of short fibres which will create problems in spinning and in dyeing because more the dead fibre content in the lot, more will be the unevenness in dyeing due to the properties of dead hair. The dead hair or guard hair do not uptake dye adequately (Fig. 22).

The results are showed in Table 6.

Table 6: Comparison between manual dehairing and mechanical dehairing

Parameter	Manually dehaired		Machanicaly dehaired	
	Mean ± SE (Range)	CV %	Mean ± SE (Range)	CV %
Fiber diameter (µ)	12.57 ± 0.64[a] (11.39 – 14.82)	19.21	12.25 ± 0.11[a] (11.08 – 13.10)	20.26
Fiber length (mm)	60.33 ± 0.21[b]	26.63	49.04 ± 0.19[a] (46.03 – 52.03)	23.22
Bundle strength (g/tex)	10.1 ± 0.11[a] (9.23 – 14.32)	17.72	9.03 ± 0.28[a] (8.08 – 13.13)	21.07
Coefficient of friction (µ)	0.61 ± 0.005[b] (0.59 – 0.64)	2.56	0.58 ± 0.007[a] (0.55 – 0.61)	3.72

The surface properties of fibre on SEM imaging showed damaged surface with fibre dehaired in the machine whereas no such damages were seen in the manual dehaired fibre. This study conclusively indicates that the machine dehairing results in reduction of strength of the fibre and damages the surface structure (Fig. 23).

Figure 23: Machine dehairing vs hand dehairing

Due to the damage of this delicate fibre, the final product quality of the end product maybe affected negatively. There is a need to develop dehairing system so that the maximum recovery is possible with minimum passage through the dehairing machine.

3.2.5   Scouring

The wool scouring is generally recommended for greasy wool to remove the grease in the fibre and other contaminants like vegetable impurities, dust, dungs, etc. Scouring process is to clean the fibres with soap, soda and water in the past times and now the use of non-ionic detergents which does not damage the fibre surface so much. Scouring at such a high level of expansive machineries probably is not the most preferred with the cottage scale knit wear industries. Such expensive machineries and equipment's are not needed in the counties like India, Nepal, Afghanistan, Pakistan and Iran because most of the Pashmina shawl making or knit wear making industries are in cottage scale which cannot afford to buy such machines.

Another reason for not going for so much of capital investment is that because it has only 5% containments including the vegetable impurities and residual grease contents. It is not in favour of having such an expansive extraction plant when the grease content is already low (Figs. 24 and 25).

Figure 24: Scouring by hands

The scouring in simple terms is to wash the fibre with non-ionic detergents to make it clean fibre before it goes to the next process, which is blending, carding, etc.

Pashmina fibre contains about 5% contaminants.

Figure 25: Scouring of pashmina fibre by machine

Traditionally in India and Nepal the scouring is not done in fibre stage, but it is carried out at yarn stage before dyeing and weaving. The scouring is done by three to four bowls scouring machine with 0.2GPL non-ionic detergents for 10min. The Pashmina fibre is also scoured using this sort of machine with lesser number of bowls, generally three to four bowls are good enough. The wool scouring being an important component of the whole process and using sophisticated machineries, it is felt necessary to discuss the wool scouring chemistry little more in details. However, it is to be kept in mind that this cottage industry scale units may go to the higher ranks in industry and they all are needed this information, rather we are preparing them for the eventuality.

The reality is that 90% of the pashmina industries are in cottage industry scale and they do not have capacity for such a huge capital investment. The big industries which are all vertical Units means where sophisticated machineries are used with very high productivity (Fig. 26).

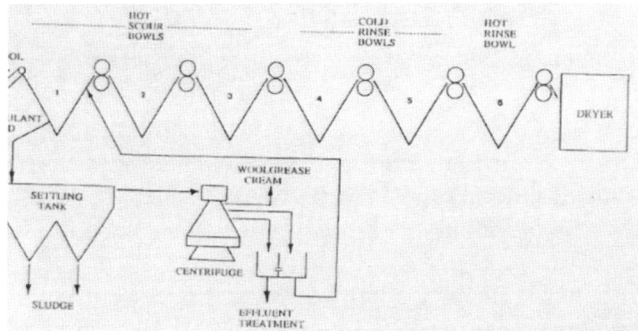

Figure 26: A simplified diagram of a typical conventional 6-bowls wool scour

The figure above is a typical commercial wool scouring plant showing the liquor treatment for dirt and grease removal. The latest plants have 8-bowls sophisticated controlled system, multiple centrifuged. The wool passes through a series of bowl, each separated from the next by large pressurise squeezed rollers. There are usually 6-bowls: the first three are scouring bowls contains hot detergent solution (above 60°C and 1–3 g/l). Next two bowls are cold rinse bowls. The squeeze rollers minimise the carryover of contamination from one bowl to the next and this is assisted by general flow of hot liquors against the direction of the wool flow. The amount of wool a single scour can process depending upon the width of the machine and the range is from 0.6 to 5 tons of greasy wool per hour. The wool passes through the whole plant scouring, drying and baling in about 20–30 min. The pashmina fibre being a delicate fibre may not be able to withstand the rigors of the process itself. That is why; the pashmina scouring is mostly done in the yarn form at the tape scouring machine that is also in the 'A' category plants which uses most sophisticated machineries and equipment's for processing the wool (Fig. 27).

Figure 27: Tape scouring machine for pashmina yarn

In the case of pashmina though, it is not applicable at all because they are processed in most of the fibre processing places in the world in very primitive way. This is to have better knowledge and understanding of the processing of animal hair which are similar in chemical structure as of wool.

Hence, the basic understanding of the machine scouring is necessary. Wool and pashmina fibre are such an example. Wool is similar to pashmina fibre and both are fibrous protein. The building blocks of proteins are of 20 amino acids, all but one which have the formula NH_3–CHR–CO_2. The making of dipeptide (NH_3–CHR–CO–NH–CHR–CO_2) as a result of various chemical changes with the condensation of another amino acid gives a tripeptide and the process continues to form a poly peptide. With 20 different R-groups, gives the protein its unique property.

Wool in reality and pashmina fibre are having complex biological structure. There is a strong relationship between the structure of the fibre and its mechanical properties.

Yarn scouring

Yarn scouring removes the back-processing lubricants along with the residual grease, dirt and dust from the yarn making operations. To scour yarn either non-ionic or less commonly, anionic detergent may be used. The most common type is nonylphenol ethoxylate type.

More recently, alcohol ethoxylates have been introduced in the use because of their easier biodegradability effluent.

3.2.6   Carbonising

The pashmina fleece may have contaminants like burrs, seeds, twigs, leaves and straw. The presence of these vegetable matters can cause series of problems during spinning, weaving, dyeing and chemical finishing (Fig.28).

Figure 28: Carbonising of pashmina fibre

Carbonising is required only if it contains impurities ≥5%. The carbonising is nothing but treatment with sulphuric acid to dissolve any vegetable impurities. This process may be affecting the surface texture of the fibre.

3.2.7   Bleaching

To prepare pale shades and pure white products the bleaching is a necessary process which uses hydrogen peroxide bleaching in mild alkaline pH.

This bleaching process improves the brightness and reduces the yellowness of the fibre. The bleaching is carried out in relatively clean liquors of the final scour bowl, at about 60°C and at a pH of 4–5 with formic acid. The use of formic acid is often not necessary and neutral bleaching is increasingly common. The chemical reaction which bleach the wool/pashmina are unknown but they take place in the dryer after application of peroxide, rather than in the bowl itself. The most obvious chemical effect of bleaching is that a portion of cysteine disulphide bonds are oxidised to cystic acid in which the sulphur is present as $-SO_3H$ group. This impairs the setting chemistry of wool/pashmina and if carried to excess can affect the mechanical properties and chemical resistance. Therefore, bleaching must be carefully controlled and confined to peroxide concentration of less than 10g/l.

3.2.8   Blending

The blending operation is done to:
- Mainly to optimise the cost and quality.
- Innovate new ideas for enhancing the products and processes.
- Deal with any fibre problem like entanglement, etc.
- Increase the production.

Figure 29: Blending of pashmina fibre

Blending of pashmina with other fibres like wool is preferred for the development of new products and better price realisation. It can also be blended with Angora goat hair, silk fibre, acrylic fibres, etc. (Fig. 29).

For example, blending with wool in the right proportion reduces the cost up to 30–40% and increases the mechanical properties of the final yarn. This is usually done after scouring and carbonisation.

3.2.9   Carding

For preparation of spinning, a process called carding is very important, let's say most important process in the whole chain of yarn making. Blended wool is carded to produce sliver which will go next to three gilling processes.

In the worsted line of spinning, carding comes after blending. Till blending, the fibres remain entangled with lots of vegetable impurities, etc. Carding is a process by which all those entangled wool and other wool fibre are processed through one big cylinder and four more small cylinders.

After cleaning up by the carding cylinder wires, it prepares all this fibre straighter and more oriented and slivers after carding goes for gilling.

The carding operator continuously check the burr rollers to judge how much burr is getting collected because that is the indication of how good the operator and the machine is. The operator uses a gauge and always checks whether it is right or not. The affectivity of the machine depends very much on how much contamination the machine removes. The feeds of the carding machine are to be very carefully done. It should not be more than prescribed feed to keep the uniformity of the carding operation. The sliver coming out of the machine is stored in the can especially made for such purposes.

Those eco-friendly processers of Pashmina Shawls who are too small and are in cottage scale level will soon be converting themselves to the bigger units with modern machineries. This is needed for their survival to stay competitive in the marketplace

3.2.10   Spinning

It is the most important operation in the whole process in making pashmina shawls, stoles and knit wear. As the experts believes that if the yarn is good, it means that 70% of the job is already done.

There are four types of spinning systems that are commonly used in wool industry and naturally for pashmina also these systems are equally applicable. China, Mongolia, Afghanistan, Australia, USA, Italy, Nepal and India – where the part of the industry is still working with the relatively primitive methods of dehairing, combing, shearing, carding and spinning. Nepal and India have the major portion belongs to cottage industry scale.

The four spinning processes are woollen system of spinning, semi-worsted, worsted spinning and hand spinning system but none of these are in use for making 100% pashmina shawls in India and Nepal, and they are still with classical hand processing: sorting, carding and spinning. The cost of the modern spinning machines is very high. They believe that hand processed yarn and hand-woven shawls are much better in quality than machine woven shawls out of machine made yarn. The prices are also set accordingly in the market place (Fig. 30).

Figure 30: Spinning of pashmina fibre

The four processes are as follows:

1. Woollen system of spinning

This process makes thicker yarn and this type of yarn are mainly used in rugs and carpet making and to some extent knit wear.

The process sequence is given in Table 7.

Table 7: Process sequence for woollen system of spinning

Scour → Carbonise → Carding → Condensing → Slubbing → Spinning → Wollen yarn

If we compare semi-worsted spinning with woollen system of spinning then woollen spun yarns are much more in demand in rugs and carpet sector. This system of spinning is easiest because of its easy processing.

2. Worsted spinning

Worsted spinning is normally used in apparel textiles. This is much expensive and takes longer time to process for production. This is probably the finest yarn spun from wool and any other similar fibre.

The difference between worsted spinning and semi-worsted spinning lies only on combing operation that means semi-worsted plus combing is equal to worsted.

The process sequence for worsted spinning is shown in Table 8.

Table 8: Process sequence for worsted spinning

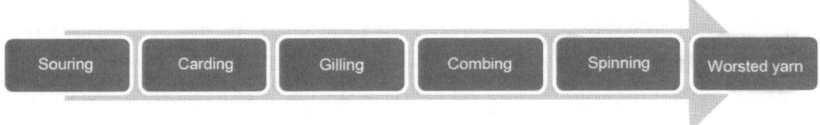

The worsted spinning system is available only in the organised sectors in various pashmina producing countries. The yarn count normally produced are 1–5 Nm in woollen system of spinning and 2–11 Nm in semi-worsted spinning. The worsted system of spinning is as fine as 150 Nm.

3. Semi-worsted spinning

Innovation and commercialisation of this system of spinning is not very old. It comes up only in eighties and since then it is doing reasonable well in the carpet sector and knit wear sector.

The system of semi-worsted spinning is different than the woollen spinning. The difference between these two-spinning systems of yarn is that one intermediate process that is gilling (sliver making). Gilling is a step to make wool more oriented, this more parallel orientation helps in achieving higher quality.

The process sequence for semi-worsted yarn spinning is given in Fig. 31.

4. Hand spinning system

The hand spun yarns are quite popular in making eco-friendly woollen rugs and carpets and normally this is a much thicker yarn having not much uniformity. The process sequence seems to be pretty long as shown in Fig. 32.

To spin yarn from 100% pashmina fibre is difficult as it is not a stronger yarn to with stand the rigour of the wool processing. That is the reason why 100% pashmina fibre is mixed with nylon to give the yarn enough strength to withstand the further processes.

The nylon fibre of the yarn is then dissolved by treating it with commercial hydrochloric acid either after spinning or after product development. In this process, dehaired and carded slivers of pashmina fibre is passed through gill boxes 3–4 times to remove the sort fibre and to parallelise them further. At this stage, the nylon fibre in sliver form is blended with pashmina fibre in the proportion of 50–50 and allowed them to undergo 5–6 passages in gill box for proper blending. The resultant sliver is then converted into roving on last pre-spinning machine that is bobbiner. The roving is then taken to ring frame for spinning. The produced yarn is double to get the required strength which is then used for product development.

Semi worsted yarn spinning flowchart

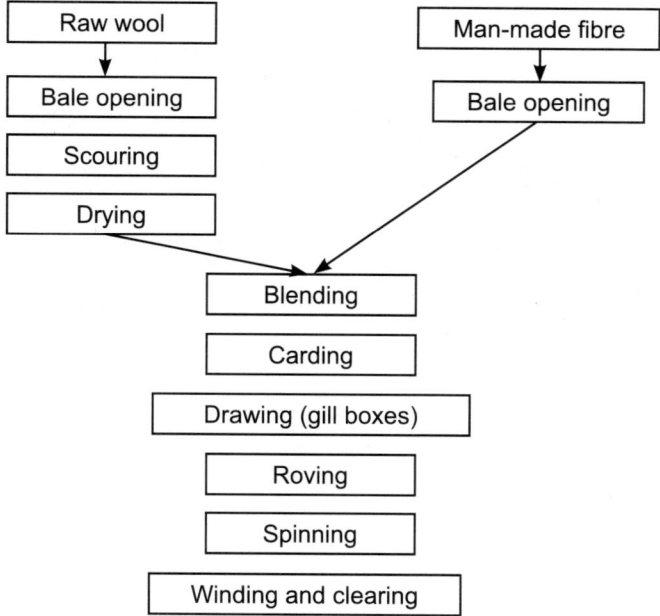

Figure 31: Semi-worsted yarn spinning

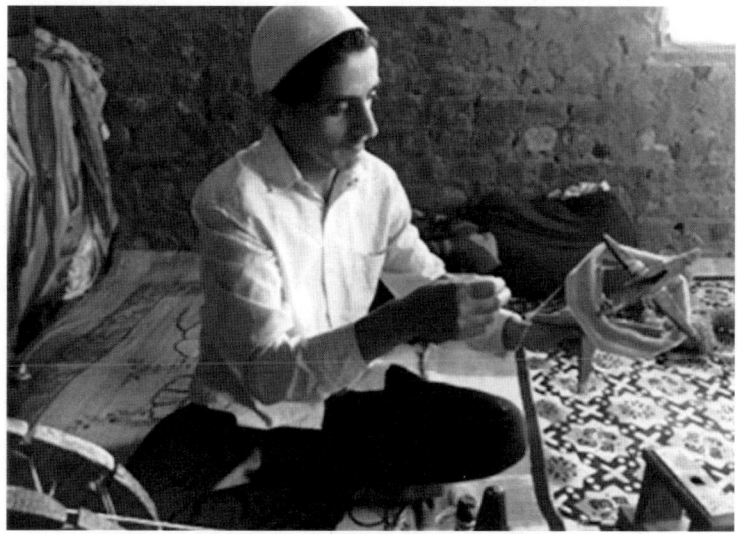

Figure 32: Hand spinning system

The blended yarns are then treated with the acid to dissolve and remove the nylon portion to get the desired 100% pashmina fabric. The results shows that the hand spun yarn made fabric are as good as traditional hand spun yarn made shawls in terms of strength, handle, feel, warmth and other performance properties. However, the abrasion loss is 50% which is the only limiting factor. One of the demerits of the process includes weakening of the fibre. The use of dissolving acid has to be just right to get the fibre not so much damaged. This is not such an eco-friendly process (Table 9).

Table 9: Performance properties of pashmina fabrics produced from hand spun and machine spun yarns

Property	Hand spun yarn	Machine spun yarn
Breaking strength (kg)	4.964	4.975
Extension (%)	40.72	25.67
Alkali solubility (%)	38.00	46.00
Abrasion loss (%)	3.75	5.73
Thermal insulation (tog)	2.00	1.95

Pashmina fibre is also spun in machine by using polyvinyl alcohol fibre (PVA) instead of nylon as carrier fibre. PVA fibres are soluble in hot water. To remove the PVA from the fabric, the fabric is treated with hot boiling water instead of hydrochloric acid. The advantage of this technique is that the surface of pashmina fibre does not get damaged and the method of spinning is considered as eco-friendly but costly because of higher price of PVA than hydrochloric acid.

These are the few important checks one should do for the whole process right from blending to the spinning for worsted yarn manufacturing and worsted fabric manufacturing.

- Roving should be stored in conditioning loom where Rh is around 30%.
- One should be careful about running two spinning frames one opposite to another, running on two different colours especially black and white. This creates heavy contaminants in both the yarns.
- One should not change the draft pinion if variation is observed in count whereas it should be controlled in preparatory and have standard feed materials.
- The ring frame should be installed with travelling over a blower on all machines.
- Temperature inside the department should not be very high because this is the input.

3.2.11   Weaving and knitting

Weaving

In weaving it is mainly the shawl weaving. Shawl is an outer wear for ladies. There are enormous number of shawl designs and structure which are available in the marketplace. The basic shawls, on which we go for various value additions are a plain woven fabric with warp and weft of all pure pashmina or in blends with fine merino wool. The shawls are woven primarily by using modern looms or the primitive pit looms. The interesting fact is that those primitive handlooms with extremely primitive way of finishing make a wonderful shawl with all the qualities which make pashmina a treasured fibre (Fig. 33).

Figure 33: The loom for weaving pashmina shawls

The handloom woven pashmina shawl has more demand than power loom woven. This is simply because the power loom woven shawl contains few pertinent defects which arise from the machine dehairing. Dyeing defects, shrinkage in the fabric and neps in the yarn are some common defect. These factors do not contribute to the overall value addition of the product. The handloom weaving is basically done by men who are highly skilled with the pashmina shawls. To make one shawl they take almost a week of size 2 m × 1 m (Fig. 34).

Figure 34: Weaving of pashmina fabric (shawl) by hand loom

The weight of the hand woven shawl is approximately 200 g. The ends and picks per inch of pashmina shawl is generally kept between 50–60 and 46–56, respectively. The dimension of the ladies, gents and stole are 2.1 m × 1m, 2.5 m × 1.37 m and 2 m × 0.8 m, respectively. The fabric weight (GSM) is kept at 50–70 g. After weaving, the fabric is hand massaged for releasing the stretches inserted during spinning and weaving.

Figure 35: Pashmina in organised sector plant

The pashmina shawl made out of machine dehaired and machine spun yarn likely to have few defects which are (Fig. 35):

- Shortening of fibre,
- Chances of adulteration,
- Dyeing defects,
- Shrinkage in the fabric,
- Neps in the yarn.

As stated earlier that the handloom woven pashmina shawl has more demand in the market than power loom woven. This is because of the European mostly preferring natural products which are made without any use of power.

Knitting

The hand spun yarn is mostly used in developing countries for pashmina fabric including China, Nepal and India and to some extent in Afghanistan for manufacturing pull overs and sweaters.

Figure 36: Knitting fabric making machine

Machine knitting preferred for products like sweaters, stoles and mufflers from 100% pashmina fibre (Fig. 36).

Table 10 given below shows an indicative cost requirement to set up a small 1000 garments per day unit.

Table 10: Approximate costing of the machineries and the equipment's needed to set up a minimum size knitting unit

Sl. no.	Type of machine	Units	Cost/unit (in USD)	Total (in USD)
1	Hand knitting machine	10	366	3,660
2	New linking machine	5	535	2,675
3	Stitching machine	5	239	1,195
4	Overlock machine	2	507	1,014
5	Chain lock machine	1	478	478
6	Button hole machine	1	3,870	3,870

Contd...

Contd...

Sl. no.	Type of machine	Units	Cost/unit (in USD)	Total (in USD)
7	Button machine	1	4,433	4,433
8	Hand presses	2	71	142
9	Scissors	5	15	75
10	Cost of packing and forwarding			17,542

Please note that this Table 10 which is just an idea and very roughly done but the core meaning remains the same.

Though these costs will vary from country to country but just for giving you an example this table is created on India based data in US dollar.

Advantages of wool/pashmina knit wear

Following are the advantages of wool/pashmina knit wear: –

- For garments shape retention,
- Fibre print – for bulk and lightness,
- For comfort – absolute fineness of the fibre,
- Softness – these fibres are extremely soft,
- Natural – the pashmina fibre is always preferable.

3.2.12   Dyeing and finishing

Dyeing

The pashmina fibre is one of the costliest fibre and their products like shawl is sold in the market as plain, shaded-ombre, beaded, embroidery and hand print shawls. Each type depends upon the major post weaving processes involved in pashmina products and there are up to 30 sub-processes depending upon the final product. Before dying, pashmina products are given a mild scour with non-ionic detergent for 30 min at 40°C, in order to remove impurities adhered during weaving. Generally dyeing is carried at a temperature just below boiling point for 1h. Pashmina fibre is exceptionally absorbent fibre and it easily and deeply picks up dyes. Local dyers prefer red, blue, yellow, orange, black, green, magenta, pink and white shade for dyeing pashmina yarn. Readily used for embroidery design (Fig. 37).

Figure 37: Dyeing of pashmina shawls

Since pashmina shawls are woven close to the face of fashioner and the colour must suit each person's skin tone. The colour must be particularly fashionable and of preferred shades. The preferable shades are purple, pale lilac to a deep violet. To get the shades dyers preferred carcinogenic free synthetic dyes. Metal complex dyes are also used for pale shades with very good fastness property. There is lots of usage of natural dyes to make the shawls or knit wear eco-friendly. The natural dyes are from pomegranate, walnut, manjeet, madder, kamala and locally available sources are preferred.

Eco-friendly dyeing

Eco-friendly dyeing of wool and pashmina fabric is the subject of the day. The various researches are done to establish that the dyeing potential of the pashmina fabric is much better with the natural dyes and bio mordant from the plant material (Fig. 38).

Figure 38: Eco-friendly dyeing

There are vast numbers of natural colours which can be obtained from nature, leaves, roots, flower, barks, fruit and stem and can be explored as dyeing sources for different fabrics. The natural dyes are clinically safer than synthetic analogues in handling and use because of non-carcinogenic and biodegradable nature. Natural dyes have gained importance due to growing environmental awareness and implementation of stringent of production and use of synthetic dyes (Table 11).

Table 11: Natural dyestuffs

Sources	Common name	Botanical name	Part used	Chemical constituent	Yield %
Walnut	Akharot	Juglons nigra	Husk	Juglone	28
Pomegranate	Anaar	Punica granatum	Rind	Tannin	15
Onion	Pyaz	Allum cepa	Skin	Flavonoids	8
Saffron	Zaffran	Crocus sativus Linn.	Flower	Flavonoids	8
SAFFRON	Oak	Gravillea robusta	Leaves	Tannin	7
Myrobalan	Hartaki	Terminalia chebula	Driedfruit	Tannin	20
Madder	Munjeet	Runia tinctorum	Root	Anthraquinone	15
Tulip	Tulip	Spthodea campanulata	Leaves	Flavonoid	9
Rohida	Rohida	Tecomella undulata	Leaves	Flavonoid	7
Lumb	Lumb	Biden pilosa	Wholeplant	-	-

The synthetic dyes are designed to resist chemical and improve the quality of the product but not friendly with the environment. The chemicals used to produce the synthetic dyes are highly toxic, carcinogenic, allergic, explosive and dangers to work with. The ingestion of water contaminated with textile dyes can cause serious damage to the human health and all other living organisms due to toxicity and mutagenicity of its components. The discharge of the highly coloured dye effluents into inland coastal water is an environmental problem of growing concern.

3.2.12.1   Vegetable dyeing

Vegetable colours are suitable for dyeing of woollen yarns and pashmina yarns beside silk jute and cotton. As per the German regulation harmful amine of the azo dyes are not eco-friendly. Those dyes which are manufactured with natural sources with plants, leaves, fruits, seeds, wood, insects, etc. are considered under vegetable dyes. Finally, eco-friendly nature of the chemical dyes is a waiting media to get the international marketing, but it is fact that chemical dyes are still under research project to declare them a 'safe-dye' which has a bad effect on human-life directly.

Dyeing process

Before we take the material for dyeing, light scouring is carried out to remove the contaminants first and secondly if there is any kempy fibre that to be sorted out. Natural vegetable dyes are suitable for natural fabrics like the fabrics of cotton, wool, silk, jute.

After scouring, the pashmina fibre is taken for mordanting and salt addition. Mordanting is done to help most of the vegetable dyes which do not fix directly on the dyeing materials (like wool, cotton, silk and jute). Dyeing material is heated to the boiling point. After mixing the metal mordant at the dye bath, his yarn/fabric is again immersed in the dye bath. Dyeing arterial is heated at boiling temperature.

Potassium aluminium sulphate is added ($SO_4 \cdot 12H_2O$), this covers only 5–20%.

Copper sulphate ($CUSO_4$)...............................1–2%.

Ferrous sulphate ($FeSO_4$)...................................1–2%.

There are about three types of mordant available in the marketplace:
1. Natural mordants
    * *Cassia fistula*
    * *Terminalia chebula*
2. Mineral mordants
    * Tartaric acid
    * Tannic acid

3. Other mordants
- Alum
- Ferrous sulphate
- Copper sulphate
- Potassium dichromate

3.2.12.2   Colour fastness of pashmina

This is one aspect of the total pashmina processing chain to have a proper standardisation and quality control of pashmina yarn and Fabric for making shawls. In the quality arena we would like to discuss two subjects which are of great importance:

1. Colour fastness and
2. Eco-friendly norms.

The cumulative colour fastness rating of the pashmina shawls:

Colour fastness to light	4 or better
Colour fastness to washing	3–4 or better

Eco-friendly norms

As per changing conditions of international markets and consumers awareness the following heads are of great importance:

- Eco-friendly application,
- ISO 9000 quality control standard,
- Environmental parameters.

The meaning of eco-friendly is to use such dyes, chemicals and processing ingredients which have no content of amines and azo group especially in those fabrics or garments which are close to human skin.

3.2.12.3   List of banned amine

4-Aminodiphenyl benzidine

4-Chloro-o-toluidene

2-Naphthylamine

0-Aminoazotoluene

2-Amino-4-nitroluene p-chlorine like

2,4-Diamonoanisol

4,4-Diaminodiphenylmethane

3,3-Dichlorobenzidine

4,4-Oxydianiline

0-Toludine

2,4-Toluylendiamine

2,4,5-trimethy lanoline

Eco-auditing

Eco-auditing is carried out in two leads:

1. Eco product auditing
2. Eco production auditing

Eco-auditing

The process is covered under following heads:

(a) Assessment of conformance of textile goods to the eco criteria
(b) Use of textile goods
(c) Pollution performance caused by their use
(d) Disposal and recyclability of used textile goods

Production audit

This process includes:

(a) Textile raw material, dyes, auxiliaries used
(b) Energy sources applications
(c) Water sources and their applications
(d) Working conditions of job and environment

Finishing

Setting of wool/pashmina yarn

Setting is a process for making the yarn set, so that the twist of the yarn does not get fuzzy. This setting is compulsory for carpet yarns as carpet pile yarn should be defined correctly to make the carpet surface nice and crispy. This is not very much in use for pashmina processing. There are four main component steps in any setting reaction of a polymeric material. In wool industry, generally three types of setting are normally used. They are:

- Autoclaving,
- Chemical setting in tape scouring machine,
- Heat setting.

Figure 39: Tape scouring machine

These all processes are not generally used by the small size (cottage scale) pashmina scarf or shawl manufacturer in any pashmina producing countries (Fig. 39).

The chemistry behind the setting process is the relaxation of the chemical stress by thiol-disulphide interchange. It is predominantly the anion from the thiol group which attacks the disulphide bond forming new disulphide bridge and another thiol anion. This is a part of the chemical stress relaxation theory (Fig. 40).

$$W_3SH \rightarrow W_3S^- + H^+$$
$$W_3SH \rightarrow W_3S - SW_2 + W_1S^-$$

Figure 40: Chemistry behind setting process

The rate of the stress relaxation reaction is proportional to the concentration of the thiolate ions and the concentration of the cysteine crosslink, the latter being essentially constant. It also depends on the pH.

Pashmina products have unique softness property along with lustre. Therefore, normally it does not require any finishing process. The addition of any type of finishing agent may spoil the unique softness of the product. However, nowadays the pashmina fabrics are produced by machine spun yarn in blends with other fibres. Such blended pashmina products are treated with thermo plastic agents like silicon softener, nano based softening agent, etc. in order to improve the handle. Generally softening finishing treatment is done after dyeing using 0.5–1.5% softener under acidic condition.

The other important finishing treatment carried out on pashmina fibre by new dyers include anti moth finishing and aroma finishing. Anti-moth chemicals such as Eulon-33 added during dyeing under acidic condition. The aroma finishing is carried out using commercially available fragments micro capsules. These microcapsules are very tiny and they get embedded into the surface of the fabric and on use of that garment, these little capsules break and starts giving fragrance. This process should be very intelligently done to absorb tiny little medical aroma capsules and anti-moth chemicals properly to release their effect on use later.

Effluent management from wet processing

Yarn wet processing is a significant contributor to the total pollution load from an integrated yarn plant. The pollution problem arises largely because of the resistance of some of the processing lubricants and detergents to biological oxidation. The chemical oxygen demand (COD) of such substances such as ethylene oxide/propylene oxide block copolymer spinning lubricants and nonylphenol surfactants is substantial. These compounds render the effluents biologically hard and pose something of a problem in some cases for treatment of the mill effluents.

3.2.13   Embroidery and design

The craft of decorating fabric and other materials using needle to apply thread or yarn is embroidery. There are possibilities in embroidery to incorporate other materials such as pearls, beads, quells and sequins. Today the embroidery is used very much on caps, hats, coats, blankets, dress, shirts, stockings, etc. Enormous number of variety of threads and yarn colours are available for embroidery (Fig. 41).

Figure 41: Embroidery work on pashmina shawls

Some of the basic techniques of stitches of earliest embroidery are chain stitch, button hole or blanket stitch, running stich, satin stich, cross stitch, etc. Those stiches remain the fundamental techniques of hand embroidery today. The art of embroidery is an art of leading swing thread on the fabric in such a way that creates possibilities of creating different types of embroideries as a developmental process. As such the development of embroidery with no changes of material and techniques can be compared with the early works to the present work and it is certainly found that the earlier embroidery works were better than of now.

The ancient Greek methodology has credited god Athena with passing down the art of embroidery with weaving leading to the famed competition between herself and the mortal Arachne. The embroidery could be done in any place in the world if the material needed for embroidery is available and using proper techniques. This situation leads us to an understanding that embroidery technique can be available from royal to general peoples. However, elaborately embroidered clothing and other house hold items are seen as wealth and status. The embroidery is an expertise job and only the experts can do this work. The development of machine embroidery and its mass production came about in stages in the industrial revolution. The earliest machine embroidery used a combination of machine looms and team women labourers embroidering the textiles by hands. This was done in France by the mid eighteenth hundred. The manufacturer of machine-made embroideries in St. Gallon in eastern Switzerland made the owners absolute rich in the latter half of nineteenth century (Fig. 42).

Figure 42: Pashmina shawl with full embroidery work

The classification of embroidery is done normally considering the nature of the waste material and the stich placement of the fabric. The main categories are free or surface embroidery, counted embroidery or needle point or canvas work. In free or surface embroidery designs are applied without regard of the weave of the underline fabric.

Chinese and Japanese embroidery are the right examples.

The embroidery can also be classified by the appearance similarity. The fabrics and the yarns used in traditional embroideries vary from place to place. Wool, linen and silk yarns had been in use for hundreds of years – both for fabric and yarn.

Lately, embroidery has gone completely computerised and hi-tech using patterns digitised with embroidery software's. The machine embroidery uses lot of rayon yarns and polyester yarns, the cotton yarns cannot withstand the machine embroidery stress.

There has also been a development for free hand machine embroidery.

The embroidery whether can be used on Pashmina shawls is to be seen first.

4

Identification and differentiation

The pashmina fibre is one of the luxury fibres among all the fibres in natural animal based or synthetic fibres of similar fibre property. Any product which is termed as pashmina product is now under doubts in the marketplace for its genuinely of pashmina content. It is an expensive fibre and an expensive resource for textile industry which makes the final product more valuable. Initially, the rich consumers could only buy pashmina but now the scenario is slightly different. The Indian consumers in the middle-class population of 200 – 300 million with the required purchasing power due to the growth of the economy of India and China, there is huge domestic market which is responsible for creating demand of pashmina fibre.

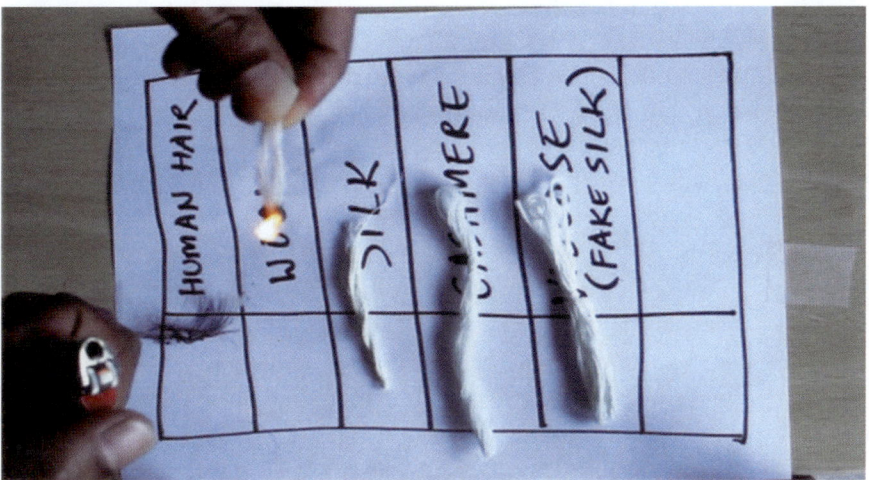

Figure 43: Differentiation of fibres

Some of the manufacturers of pashmina shawls and knit wear are so unscrupulous that they started blending pashmina fibre with another animal fibre mainly wool. This way the whole industry loses its credibility and the faith of the consumer and this resulted in making the whole pashmina/cashmere business distorted. Consumers who have to pay very heavy price for the product for which they do not know even whether they are getting 100%

pashmina or blend with some other fibre. Naturally, now these consumers will also try to stay away from such products because such products are not of essential category, probably its priority is last (Fig. 43).

The obvious choice for all the stake holders for pashmina business pipeline is to get a test method suitable for this job of identifying pashmina with other animal fibres. This job found not to be as easy as the morphological structure of pashmina is similar to wool and testing method like use of microscope cannot also identify objectively and for sure. That is the reason why we are even now could not find a solution of this identification problem.

The main test methods for identifications are expensive and time consuming. However, they cannot give guarantee for 100% correct results.

The standard test methods are:

1. Microscopy

Like microscopy technique scanning electrical microscopy (SEM) may permit accurate fibre and fibre blend identification. It does require a high degree of skill and experience. SEM analysis is difficult and time consuming. However, differentiation in between fine wool and pashmina hair can be judged almost accurately (Fig. 44).

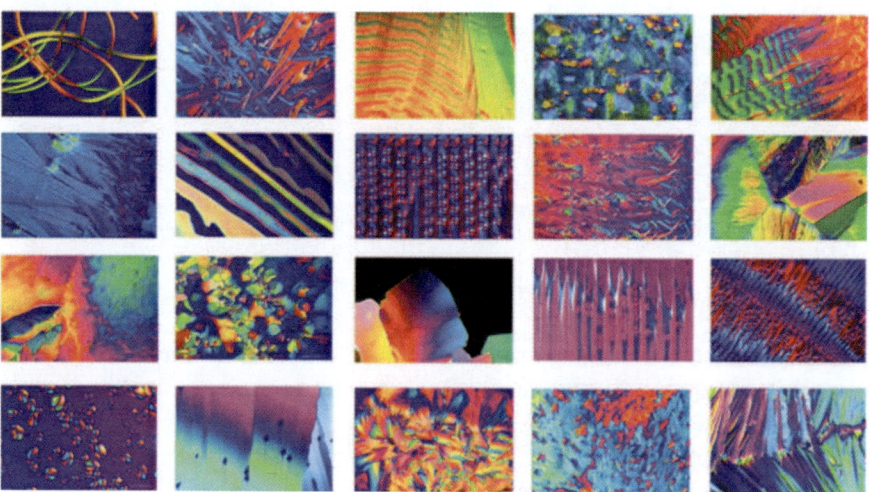

Figure 44: Electron microscopy of different pashmina fibres

SEM analysis reveals that each hair fibre has its own characteristic scale pattern. Pashmina fibre has surface scale on its structure like the sheep wool however the number of scales per micron is lesser then wool (Fig. 45).

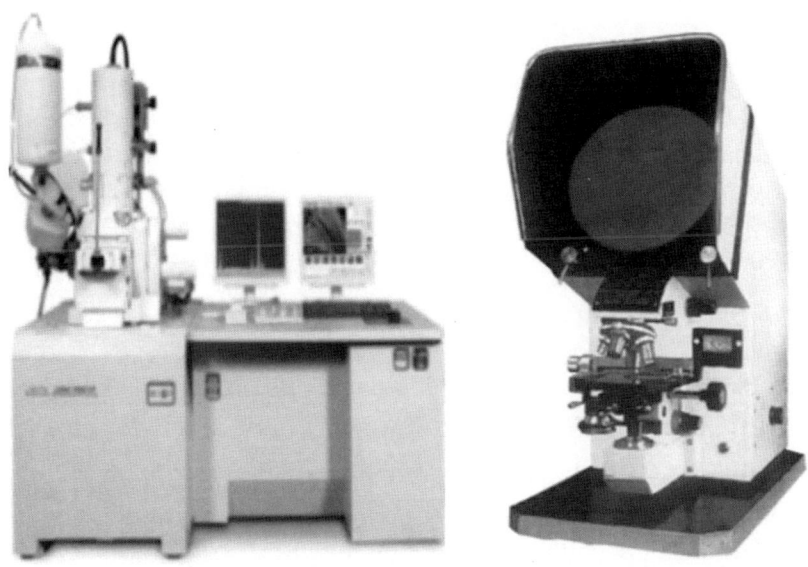

Electric microscope   Projection microscope

Figure 45

There is significant contribution in identification of animal hair fibre. But the technique is not recommended.

2. DNA sequencing

DNA polymorphism is like restriction fragment length polymorphism (PCR – RFLP) technique is used to differentiate pashmina/cashmere and fine wool fibre. The presence of DNA in animal hair shafts has enabled the isolation of DNA from cashmere/pashmina and fine wool fibre. DNA analysis can be able to distinguish between these fibres in the raw stage. But not if they have went under any heat or chemical treatment (Subramanium et al., 2005; Mccarthy, 1990).

In the textile industrial change cashmere is often adulterated with sheep wool for material gains. The quantitative analysis of cashmere-wool blends remains difficult by traditional analysis method. Scanning electron microscopy

(SEM) is the only established method for identification of fibre according to an international standard. However, it is expensive, time consuming and fails if cashmere is adulterated with descaled or stretch tool or from new breeds of cashmere goats. DNA exhibits a relatively strong capacity to resist physical and chemical treatments, this technology may work for the dyed fibre as well.

3. PAGE method

In conjunction with transmission electron microscopy and liquid analysis, 2D-PAGE (two-dimensional polyacrylamide gel electrophoresis technique is used and an alkaline gel in the first dimension and SDS in the second dimension. It also can distinguish pashmina fibre from other.

The amount of sterol and fatty acid extracted using soxhlet extraction from different animal fibres can be useful additional procedure to the conventional method for distinguishing. But this procedure is also not practical because it is expensive and time consuming.

4. Staining method

Staining with phosphotungstic acid followed by magnification of wool/pashmina fibre in transmission electron miscopy is used to identify difference between pashmina fibre and wool fibre.

Fine wool has bilateral structure of ortho and para cortex, while pashmina fibre has a range of structures bilateral to non-bilateral. Para-cortical cells are stained as black by tungsten ion that can be used to identify the fibre.

There are many suggested test methods which are also being practiced as a tool for checking pashmina fibre content in yarns and shawl. Unfortunately, none of the existing and recommended test method is useful, which should be in a competitive cost and takes less time. Some suggested testing are:

(a) Scale height – which is a useful criterion for distinguishing pashmina.
(b) Estimation of cysteine and cystic acid contents – is also a means for identifying pashmina from any other pashmina hairs.
(c) Alkaline solubility, urea-bisulphite solubility, effects of acid, alkali and enzymes of wool and pashmina fibre.
(d) Plasma etching and scanning electron microscopy.
(e) A technique for understanding the internal structure of keratin fibres.

Inspite of all such test methods which are available now, none of them are exactly suitable for the purpose. Therefore, its need of the time to develop a method by joining hands with any reputed educational institute or research organisation to develop a method which can differentiate pashmina fibre easily, in less price and time.

The noble microchip real time PCR may prove a valuable technology to identify and quantifying cashmere in wool content in the fibre mixture by wool adulteration of cashmere. The data on cashmere samples procure from different region and from cashmere goats of different age supports the validity of technology. The test is very cost effective, highly specific and sensitive with minimised human error and minimal false negative or false positive rates. This method is based on testing mtDNA extracted from the cashmere and wool can be used for mass testing and for this purpose, the ready to run microchip may prove user-friendly to the lab technician.

5

Standardisation and quality of pashmina

Expected quality of pashmina by hand made yarn and machine-made yarn are both to be checked for its characteristics. Pashmina is known as the most expensive fibre in the world, hence, quality control at the manufacturing stages is as important as the quality and inspection of the end product. The quality to be checked at every level of productions is given in Table 12 between hand spun and machine spun yarn.

Table 12: Yarn parameters of pashmina shawls

Characteristics	Hand spun	Machine spun	Blended yarn
Count (Nm)	702	90/2	55/2
Yarn strength	70 + 10	30 + 10	140 around
Elongation (%)	4% + 1	2% + 1	5%
U %	22 + 2	30 + 2	18 + 2

The related detail specification from International Wool Textile Organisation

IWTO RED BOOK

SPECIFICATIONS

Edition: 2013 / 2014
Date of Issue: 28.10.2013

Published by:
International Wool Textile Organisation
Copyright:
© International Wool Textile Organisation 2013
Any content printed or downloaded may not be sold, licensed, transferred, copied or reproduced in whole or in part in any manner or in or on any media to any person without the prior written consent of the International Wool Textile Organisation.

INTERNATIONAL
WOOL TEXTILE
ORGANISATION IWTO

INDEX of IWTO SPECIFICATIONS: Test Methods and Draft Test Methods   Edition: 2013-2014

IWTO-0-12:   Introduction to IWTO Specifications. Procedures for the Development, Review, Progression or Relegation of IWTO Test Methods and Draft Test Methods

App A (2013):   IWTO Technology & Standards Committee Organisational Chart
App B (2012):   Presentation of Supporting Technical Data
App C (2012):   Guidelines for the Presentation of IWTO Specifications
App D (2009):   Statistical Methods
App E (2012):   Working Group Drafts

IWTO-2-2007:   Method for the Determination of the pH Value of a Water Extract of Wool

IWTO-3-1986:   Method of Test for the Acid Content of Wool

IWTO-6-2013:   Method of Test for the Determination of the Mean Diameter of Wool Fibres in Combed Sliver using the Airflow Apparatus

IWTO-7-2011:   Sub-sampling Staples from Grab Samples

IWTO-8-2011:   Method of Determining Fibre Diameter Distribution Parameters and Percentage of Medullated Fibres in Wool and other Animal Fibres by the Projection Microscope

IWTO-10-2003:   Method for the Determination of Dichloromethane Soluble Matter in Combed Wool and Commercially Scoured or Carbonised Wool

IWTO-12-2012:   Measurement of the Mean and Distribution of Fibre Diameter Using the Sirolan-Laserscan Fibre Diameter Analyser

IWTO-17-2011:   Determination of Fibre Length and Distribution Parameters

IWTO-18-2000:   Method for the Determination of Evenness of Textile Strands using Capacitance Testing Equipment

IWTO-19-2012:   Determination of Wool Base and Vegetable Matter Base of Core Samples of Raw Wool

IWTO-20-2004:   Method for the Determination of the Felting Properties of Loose Wool and Top

IWTO-26-2004:   Glossary of Terms Relating to Sampling

IWTO-28-2013:   Determination by the Airflow Method of the Mean Fibre Diameter of Core Samples of Raw Wool

IWTO-29-2003:   Method for the Determination of the Dimensional Change induced by Free-Steam in Fabrics Containing Wool

IWTO-30-2007:   Determination of Staple Length and Staple Strength

IWTO-31-2002:   Calculation of IWTO Combined Certificates for Deliveries of Raw Wool

IWTO-32-2005:   Determination of the Bundle Strength of Wool Fibres

IWTO-33-2003:   Method for the Determination of Oven-Dry Mass and Calculated Invoice Mass of Scoured or Carbonised Wool

IWTO-34-1998:   Determination of Oven-Dry Mass, Calculated Invoice Mass and Calculated Merchantable Mass of Wool Tops

IWTO-35-2003:   Method for the Measurement of Colour of Sliver

IWTO-38-1999:   Method of Grab Sampling Greasy Wool from Bales

IWTO-41-1992:   Determination of the Invoice Mass of Scoured or Carbonised Wool or Tops or Noils by Capacitance Method

IWTO-42-2002:   Crease Pressing Performance Test

IWTO-47-2013:   Measurement of the Mean and Distribution of Fibre Diameter of Wool using an Optical Fibre Diameter Analyser (OFDA)

IWTO-49-2005:   Formability Test

IWTO-50-1994:   The Measurement of Dimensional Stability and Hygral Change in Woven Fabrics

IWTO-51-1994:   Measurement of the Stability of Surface Finish on Woven Wool Fabric (amended 1994)

IWTO-52-2006:   Conditioning Procedures for Testing Textiles

IWTO-55-1999:   Method of Automatic Counting and Classifying Cleanliness Faults in Tops Using the Optalyser Instrument

IWTO-56-2013:   Method for the Measurement of Colour of Raw Wool

IWTO-57-2000:   Determination of Medullated Fibre Content of Wool and Mohair Samples by Opacity Measurements using an OFDA

IWTO-58-2000:   Scanning Electron Microscopic Analysis of Speciality Fibres and Sheep's Wool and their Blends

IWTO-62-2010:   Determination Of Fibre Length, Length Distribution, Mean Fibre Diameter And Fibre Diameter Distribution Of Wool Top & Slivers By The OFDA4000

IWTO -65-2013:   Determination of Pilling and Fuzzing of Wool and Cashmere Knitted Fabrics Using the Pill Box

DRAFT TEST METHODS

The main difference between an IWTO Test Method and a Draft Test Method is that the latter has not yet demonstrated sufficient reproducibility to meet the technical standards for acceptable inter-laboratory variation. Whilst Draft Test Methods define the standard methodology being developed, they have no official status for commercial usage, unless agreed between the contracting parties.

Draft Test Methods represent the first formal approval stage in the development of IWTO Test Methods. They provide an opportunity for both technical and commercial evaluation of the developing methodology, during its logical progression to full standardisation.

Under normal circumstances, a developing Specification will remain at Draft Test Method status for a minimum of 2 years, to provide a reasonable period for its applications to be examined and its commercial implications to be understood.

In special instances, such as when demonstrable weaknesses have been found, a full Test Method may be downgraded to Draft Test Method status until its weaknesses have been satisfactorily addressed or until it is downgraded further to Working Group Draft.

DRAFT TM-1-2002:   Method of Determining "Barbe" and "Hauteur" for Wool Fibres Using a Comb Sorter

DRAFT TM-4-2000:   Method of Test for Determining the Solubility of Wool in Alkali

DRAFT TM-5-1997:   Method of Determining Wool Fibre Length Distribution of Fibres from Yarns or Fabrics Using a Single Fibre Length Measuring Machine

DRAFT TM-13-1997:   Counting of Coloured Fibres in Tops by the Balanced Illumination Method

DRAFT TM-16-2002:   Method of Test for Wool Fibre Length using a WIRA Fibre Diagram Machine

DRAFT TM-24-2001:   General and Specific Methods for the Determination of Cleanliness Faults in Combed Wool Slivers

Appendix 2 Supplement 3: Counting of Straws, Bast Fibres and Burrs Greater than 10 mm

DRAFT TM-40-2002:   Determination of the Abrasion Resistance of Wool and Blended Wool Fabrics using a Martindale Machine

DRAFT TM-43-1998:	Measurement of Solvent Extractables for Scoured Wool or Sliver Using Near Infrared Analysis
DRAFT TM-45-1999:	Determination of Cashmere Down Yield for Core Samples of Cashmere Fibre
DRAFT TM-59-2009:	Method for the Determination of Chemical Residues on Greasy Wool
DRAFT TM-60-2001:	Method for the Measurement of Fibre End Characteristics in Wool Slivers as a Guide to Fabric Skin Comfort
DRAFT TM-61-2001:	Method for the Determination of Petroleum Ether Extractable Matter in Wool Yarns and Certain Wool Blends
DRAFT TM-63-2007:	Determination of the Invoice Mass of Tops, Noils, Scoured or Carbonised Wools by the Malcam Microwave Method
DRAFT TM-64-2012:	Method for the Fibregen Sliver Cleanliness Testing System

WORKING GROUP DRAFTS

DRAFT TM-9-97:	Method of Test and Assessment for Proofness of Wool Fabrics against the Common (Webbing) Clothes Moth
DRAFT TM-11-99:	Method of Test for the Solubility of Wool in Urea-Bisulphite Solution
DRAFT TM-14-97:	Method of Test and Assessment for Proofness of Wool Fabrics against the Black Carpet Beetle
DRAFT TM-15-98:	Method for the Colorimetric Determination of Cystine Plus Cysteine in Wool Hydrolysates
DRAFT TM-21-99:	Method for the Determination of the Alkali Content of Wool
DRAFT TM-22-02:	Method for the Determination of the Weight per Unit Area of Woven Cloth
DRAFT TM-37-02:	Determination of Crimp of Yarn in Fabric Containing Wool
DRAFT TM-39-02:	Determination of the Number of Threads per Centimetre in Woven Fabrics

The fabric quality parameters are different for different types of pashmina shawls. Table 13 shows the general characteristics of pashmina shawls.

Table 13: Fabric quality norms for different pashmina shawls

Characteristics	Requirement
Length	As declared + 2 cm
Width	As declared + 2 cm
Pashmina hand spun mass (g/m)	70+5
Pashmina machine spun mass (g/m)	60+5
Pashmina blended shawl mass (g/m)	110+5
Pilling resistance	4 or better
Colour fastness to light	4 or better
Colour fastness to washing	3–4 or better
Staining on adjacent fabric	3–4 or better

The problem with cottage scale industries is that all the processes are done by hands and obviously the quality between shawls to shawl differs.

The end product – 100% pashmina shawl is such an excellent product with its softness, handle and feel. Sometimes the actual quality standards are over looked.

The following are the quality criteria up to yarn stage which has to be maintained or sustained as a part of the standardisation:

- Fibre diameter (in micronaire value),
- Fibre staple length,
- Fibre strength,
- Uster values to be checked,
- Dyed yarn matching with the master sample,
- Natural colour of raw pashmina,
- Fastness tests for dyed pashmina.

Figures 46 and 47 which are taken as example showing how the pashmina top analysis results from an organised sector company. This company is in the organised sector having used the most modern equipment available in this field.

Figure 46: Pashmina tops results

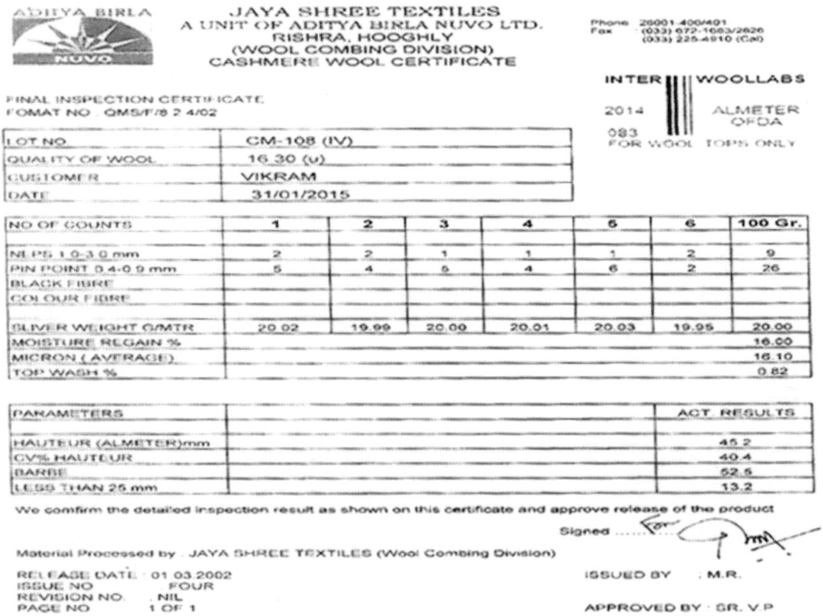

JAYA SHREE TEXTILES
A UNIT OF ADITYA BIRLA NUVO LTD.
RISHRA, HOOGHLY
(WOOL COMBING DIVISION)
CASHMERE WOOL CERTIFICATE

Phone 26001 400/401
Fax (033) 672-1063/2626
(033) 225-4910 (Cal)

INTER WOOLLABS
2014 ALMETER OFDA
083
FOR WOOL TOPS ONLY

FINAL INSPECTION CERTIFICATE
FOMAT NO . QMS/F/8 2 4/02

LOT NO	CM-108 (IV)
QUALITY OF WOOL	16.30 (u)
CUSTOMER	VIKRAM
DATE	31/01/2015

NO OF COUNTS	1	2	3	4	5	6	100 Gr.
NEPS 1.0-3.0 mm	2	2	1	1	1	2	9
PIN POINT 0.4-0.9 mm	5	4	5	4	6	2	26
BLACK FIBRE							
COLOUR FIBRE							
SLIVER WEIGHT G/MTR	20.02	19.99	20.00	20.01	20.03	19.95	20.00
MOISTURE REGAIN %							16.00
MICRON (AVERAGE)							16.10
TOP WASH %							0.82

PARAMETERS		ACT. RESULTS
HAUTEUR (ALMETER)mm		45.2
CV% HAUTEUR		40.4
BARBE		52.5
LESS THAN 25 mm		13.2

We comfirm the detailed inspection result as shown on this certificate and approve release of the product

Signed ......

Material Processed by . JAYA SHREE TEXTILES (Wool Combing Division)

RELEASE DATE : 01.03.2002
ISSUE NO FOUR
REVISION NO. . NIL
PAGE NO 1 OF 1

ISSUED BY : M.R.

APPROVED BY : SR. V.P

Figure 47: CM 108 top results

Table 14: Cashmere tops summary results

Colour	H	CVH%	<30 mm	μ	μ cv%	Test No.	Date of Test	Test conducted by
Ivory	46.5	37.3	19.10	16.51	21.5	HKSL1403039941TX	Mar 22, 2014 – Mar 27, 2014	SGS
						TSNT00616107	06-Mar-14	Intertek
Ivory	46.9	37.4	18.20	16.51	22.4	TSNT00631385	23-Apr-14	Intertek
White	42.3	42.5	29.10	15.5	22.6	TSNT00638055	07-May-14	Intertek
Brown	47.6	36.9	17.80	16.53	23.4	TSNT00648587	06-Jun-14	Intertek
Ivory	47.1	39.5	19.50	16.63	22.2	TSNT00648587	06-Jun-14	Intertek
Ivory	46.7	40.5	20.20	16.5	25.2	TSNT00689475	10-Nov-14	Intertek
Ivory	47.6	39.7	19.70	16.6	23.9	TSNT00689475	10-Nov-14	Intertek
Ivory	47.9	39.8	19.30	16.6	24.4	TSNT00705852	1.12.2014	Intertek
White	51	41.6	17.60	16.67	23.0	TSNT00697052	11-Nov-14	Intertek
White	50.2	41.9	18.0	15.5	22.8	TSNT00716588	12-Jan-15	Intertek
White	50.6	41.3	17.5	16.53	26.2	TSNT00717265	14/1/2015	Intertek
Ivory	45.2	40.4	23.1	16.1	23.5	TSNT00717286	14/1/2015	Intertek

Yarn pre-fabric operation and final fabric are the three most important criteria to be checked. For pashmina fabric following are the criteria to be checked (Table 14, Fig. 48):

- Fabric weave structure
- Fabric dimension stability
- Subjective checking of surface of the fabric
- Fabric durability and drapability
- Similarly, for knitted sweaters or shawls for their structure
- Knitting faults

Figure 48: Flow chart for cashmere top

6

Innovation of pashmina

Product quality and innovation are linked together. Today's quality is tomorrow's innovation. The quality is an important component of innovation and the innovation is the end process.

There are tremendous opportunities for further innovation and development because the industries are in its cottage scale level. Innovation probably is a combination of

Innovation = Invention + Market adaptability.

This formula is true for organised sectors of spinning and weaving industries where the availability of the modern machines and equipment are available.

If that is the case, we have a lot of opportunity for innovations because the type of machineries and equipment's they use today to catch up with market demand are not of standard. However, we see some potentiality in the following areas of the 100% shawl making (Fig. 49):

- The innovation in household charka (Fig.50),
- Loom mechanical upgradation,
- The dyeing process needs improvement,
- Printed design,
- Easy care treatment,
- Eco-friendly.

Figure 49: Innovation in pashmina shawls

However, the market adaptability is an everyday changing phenomenon and manufacturer has to keep up with these tremendous changes.

Innovation like "shower suit" by wool innovation of Australia and like "sleep better with wool" and Optim fibre. These are only couple of names suggested but there are many such innovations at every step of the wool processing.

Figure 50: Innovative charkha

Pashmina Info Brochure – Royal Bhaktapur suggests that Pashmina shawls are timeless, fashionable and luxuries (Fig. 51).

Figure 51: The convertion of traditional shawl into fashionable stole

A concept of innovation suggested by Royal Bhaktapur of Nepal which is quite unique in its conception. This shows how communities can join for a pashmina project and bring new ways of processing pashmina to make beautiful shawls (Fig. 52).

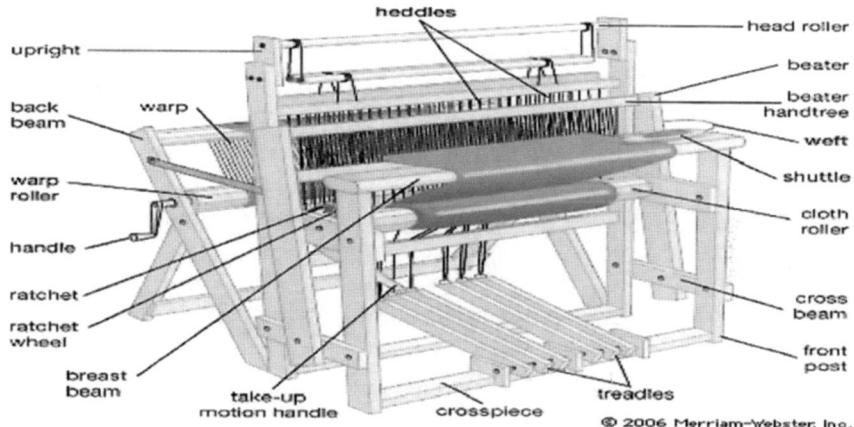

Figure 52: Innovative improvised loom

7

Future scope of research and development

As stated earlier also the future scope of improvement with modern machineries and equipment are tremendous. It will then be a challenge to the cottage scale handmade shawl makers who consider that handmade shawls cannot be replaced by machine made shawls as they are unique creation and they do not have the desire to go for machine application in their chain of processing. Modernisation of the machineries, productivity and the quality of Pashmina shawls and knit wear are the answer to compete in the global market. Research, development and innovation go hand in hand and there is no time limit of this activity, this is continuing activity as the consumers demands are changing every day. We must abide by the consumers' choice; this is called market adaptability. The marketing section of the manufacturing set up is vitally important. Study of the market both domestic and export. They are the eyes for the innovation. An innovation partly done by IWS – International Wool Secretariat to make Pashmina shawl machines washable. Unfortunately, this job was not done fully as per the manufacturer's opinion whose allegation is that Wool Mark did not complete the job as this process of super wash treatment that does not cover all types of shawls. The opinion of the Woolmark Company is that some shawls with open weaving structure will not be possible to get this treatment done. This is a natural phenomenon.

However, the company has committed the further continuation of this project to find a solution to make the pashmina shawl fabric as an easy-care product. If shawls of any type (weave) cannot be treated the same way all the types of pashmina shawls. For the shawl weave construction is a determinant factor in this treatment.

There are few jacquard and dobby designs of shawls which create problems due to their comparatively loose weave structure in completion of this process.

Few good innovations are seen which may be of use for value adding pashmina products also. In the field of pashmina dyeing and finishing the processes are like:

- Nano finishing,

- Plasma treatment,
- Enzyme treatment,
- Carbon footprint,
- Optim – the wonder fibre.

7.1   Nano Finishing

In 1974, Professor Narlo Taniguchi did observe that nano technology consists of the process of separation, consolidation and deformation. Nano technology however is the design characterisation production and application structure. Devices and system by controlling the shape and size at the nano scale. This technology works at the molecular level, atom by atom to create large structure with improved molecular orientation. Nano technology is upgrading the existing function and performance of textile material, developing intelligent textile with completely new characteristics and functions (Fig. 53).

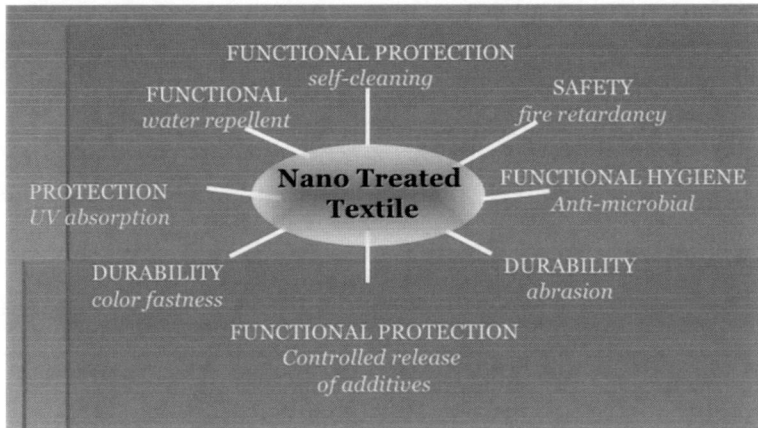

Figure 53: Nano treatment for textiles

The top down approach and the bottom up approach involving the nano particles also made by building atom by atom.

The impact of nano technology in the textile finishing process has brought up the innovative finishing as well as the new application technique. It is very well known that atoms and molecules possess different behaviour then those of bulk materials. The appeals based on nano technology are becoming very popular. These finishes with the nano technology always be carefree fabrics that minimise strength, superior liquid repellence and provide wrinkle resistance.

7.1.1   Antimicrobial finish

The growth of bacteria and microorganisms in food and water prevented when stored in silver vessels due to its anti-bacterial properties. The antibacterial properties of silver are now scientifically recognised. Silver ions have broad spectrum of anti-microbial activities (Fig. 54).

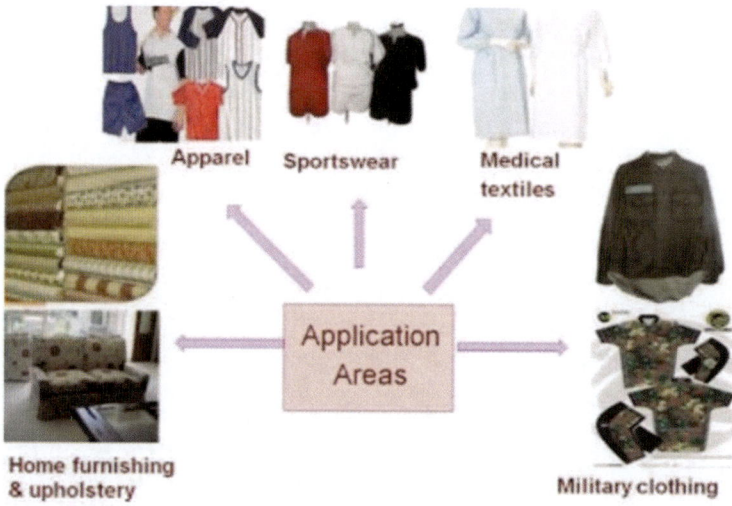

Figure 54: Application of antimicrobial finishing (Source – SMITA Research Lab)

The method of producing durable silver containing anti-microbial finish is to encapsulate the silver compound or nano particles. These microcapsules can be prepared by a two-step processing. The first step is an emulsified solution of a perfume is encapsulated with the melanin pre-condenser. In second step microcapsules so produced is treated with silver nano particles dispersed in water soluble styrene, maleic anhydrite polymer solution before it is fully died.

7.1.2   Self-cleaning suits

What a hustle it is to send your expensive suit to the underdeveloped laundry nearest to home, was the only solution to keep your suits clean. So, it is not only the inferior quality of dry cleaning, the nano technology had shown some nano paths to develop them further.

7.2   Plasma treatment

The plasma is an ionised gas with equal density of positive and negative charges which exists over an extremely wide range of temperature and pressure (Fig. 55).

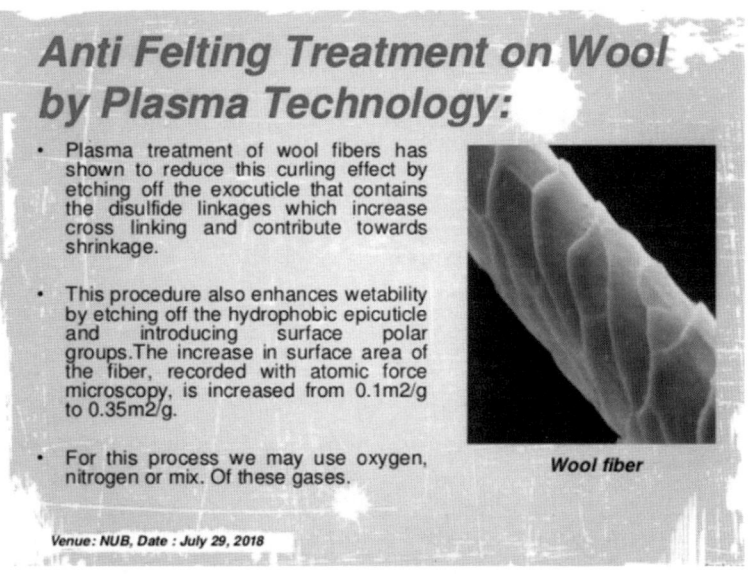

Figure 55: Plasma technology in textile (Source – slideshare.com)

The plasma consists of free electrons, ions, radicals, UV radiations and other partials depending upon the gas used. The plasma gas particles etch on the fabric surface in nano scale to modify the functional properties of the fabric. There is different methodology to induce the ionisation of plasma gas for textile treatment.

As for example:

    (a)   Glow discharge method.

    (b)   Corona discharge method.

    (c)   Dielectric barrier discharge method.

Low temperature plasma technology including both glow discharge under reduced pressure as well as dielectric barrier discharge under normal pressure have been well established in textile materials also.

The functionality of the textile materials can be improved by plasma technology:

1. Wettability.

2. Hydrophobic finishing.

3. Adhesion – plasma technology.

4. Product quality.

5. Product functionality in textile material.

It says that this plasma technology is applicable to the most of textile material including pashmina fibre for surface treatment.

7.3   Enzyme treatment

Enzymes can be used instead of chlorine bleach for removing stains on cloth. The use of enzymes also allows the level of surfactant to be reduced and permit the cleaning of clothes in the absence of phosphate (Fig. 56).

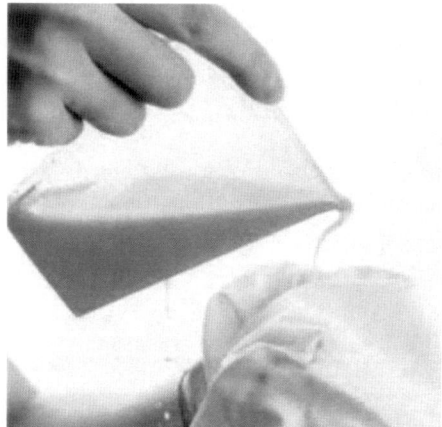

Figure 56: Enzyme treatment for removal of stains

Enzymes also contribute to safer working conditions through elimination of chemical treatment during production process.

Environmental benefits

1. Lower discharge of chemicals and wastewater and decreased handling of hazardous chemicals for textile workers.

2. Consumer's benefits.

3. Improved fabric quality.

4. Stone washed jeans without stones.

7.4   Carbon footprint

7.4.1   Introduction

The main contributor to the global warming is Co_2 i.e. carbon dioxide. The carbon footprint reveals that how much CO_2 in total is emitted around the value chain of a product. Fabrics unbelievable have a large carbon foot print that means it takes a lot of energy to produce fabrics. In the developing world where the textile industries represents larger percentage of GDP and mills are often antiquated, the CO_2 emission are greater (Fig. 57).

Figure 57: Textile and clothing industries effects on the environment

In fact, textile industry is one of the biggest sources of greenhouse gasses on earth due to huge size and scope of the industry as well as several processes and products that go into the making textiles. These facts may not be needed today but in future we may have to take care of this menace even in pashmina industry.

7.4.2   Objective

The purpose of participating in product carbon foot printing is to certify in-house assessments of product-level carbon emissions and to provide customers with independent credible verification to prove it.

7.4.3   Goals and scope of coverage

The goal of the work is to find the complete and accurate value of carbon emissions from all the unit processes involved in the production of pashmina shawl. The objective of the Life Cycle Inventory (LIC) stage of the work is to provide complete information about all the inputs and outputs in the form of elementary flow to and from the unit processes involved in the study.

7.4.4   Emission inventory for pashmina shawl production

Approximately 25% of carbon emissions in total manufacturing process of one pashmina shawl come from dyeing and packing, whereas 57% come from spinning and weaving unit.

7.4.5   Potential sources of carbon emission

The various activities related to the production process of the pashmina shawls which are the potential source of carbon emissions are (Fig.58):

- Off – site activities
(a) Rayon staple fibre production (embodied carbon footprint).
(b) Transportation of raw material (rayon staple fibre) and fuel (diesel/ petrol) to the production site and produced pashmina shawl from the production site to the warehouse (transportation carbon footprint).

- On – site activities

Easy inclusion of consumable fuel and calculation of site emissions.

Figure 58: Symbol of carbon footprint

7.5   Optim – the wonder fibre

CSIRO in Australia has taken a major innovative step to create a new fibre which will be similar or nearly like the most expensive natural fibre in the world – cashmere/pashmina fibre. This creation is known as Optim fibre – a whole new textile fibre based on wool. Optim processing technology reengineered wool fibres by stretching to make them softer, stronger and lighter than the untreated wool.

Two new types have been created – Optim fine and Optim max.

The Optim fine manufacturing process stretches 19-micron wool fibre between 40% and 55%, making on average fibre 3–3.5 micron finer. The reduced micron fibre is then chemically set in this finer and softer form. Optim fine has increased its length and strength as well as a silk like lustre, but retains many of the desirable properties of fine wool. Synchrotron analysis by X-ray deflection method conformed that the structure of the Optim fine fibre was silk like after the treatment.

OPTiM

The production of Optim max fibre, wool fibre is stretch to give an average fibre extension to about 20–30%, which is temporarily set. The stretch fibre is then blinded with normal wool and spun into yarn. During the finishing of the yarn it is immersed in hot water causing the Optim max fibre to retract to their original length. This causes normal wool fibre to buckle to distort giving a soft, light weight and bulky yarn for the manufacture of light weight garments, particularly knitwear's.

The technology was invented by David Phillips at CSIRO textile and fibre technology division.

The Optim fibre as described has a feel and handle very similar to pashmina whereas it is four times cheaper than pashmina. It means that you are getting raw material at such a lower price which is at par for quality, handle and feel of pashmina.

Unfortunately, for some reasons not very known to us, the Optim fibre processing and its end product never become popular in the pashmina/ cashmere market (Fig. 59).

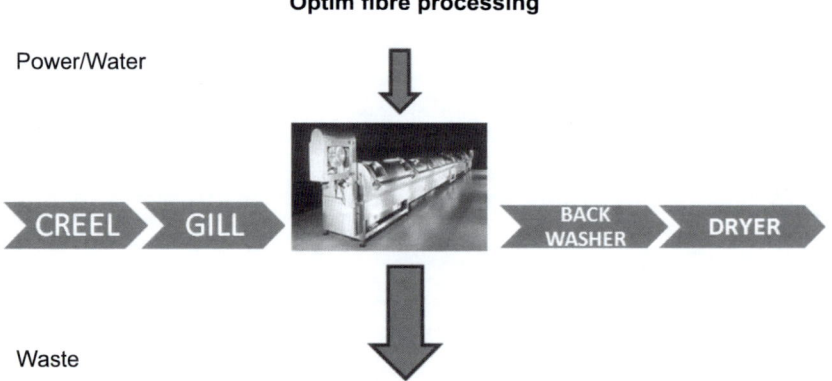

Optim fibre processing

Power/Water

CREEL ⟩ GILL ⟩   ⟩ BACK WASHER ⟩ DRYER

Waste

Figure 59: Processing sequence for Optim fibre

8

World market for pashmina and distribution channels

There are two distantly separate market exists in this world of pashmina. The medium – low end market segments where China clearly dominates and the high-end market of western European Brands are dominating. In between there is a medium segment where China, Mongolia, India and the Nepal are the major players. The world has lost a very good business for shawls, stoles and knit wears made from a fibre pashmina which is extremely soft in handle, touch and feel and the very look of it. These properties of the pashmina products historically attracted Josephine, wife of Napoleon and the British Raj in total. These pashmina shawls, scarfs and knit wears where it was a fibre of excellence, even today 100% Pashmina shawls are so unique and meant for only rich peoples of the world.

Unfortunately, pashmina – a fantastic fibre lost its image completely worldwide due to some unscrupulous manufacturer and merchandiser created fake pashmina shawls by creating blend of pashmina with lower priced synthetic fibres or maybe with finer Marino wool. Though finer wools are somewhat like pashmina is considered as acceptable because both are of same chemical composition but not accepted in the marketplace if blended with synthetic fibres. These unscrupulous peoples may have earned a lot using this wonderful natural fibre, but this made the whole world scary of buying 100% pashmina products from the market place. Pashmina is very expensive fibre and naturally the end products are also very expensive, no buyer will be comfortable for giving lot of money for purchasing fake pashmina product with the bold label saying that "it is 100% pashmina fibre".

In spite of all these negative factors against pashmina or cashmere fibre still considered as the king of luxury fibres. The demand and supply ratio of pashmina products is extraordinary (Fig. 60).

There is a much more overall demand for pashmina even today creating the imbalance between the demand and supply.

Supply in the hand of more cottage industry scale then in the composite vertical units.

Figure 60: Supply vs demand

Todays CONSUMERS

Figure 61: What consumers want?

As such today's consumers are completely different than 10 years back. Today's customers are much more intelligent, much sportier and they know how and what to pay for (Fig. 61).

Consumers today want the value for the money they have spent. They are much more relaxed in there dressing with sufficiently comfort and with easy care are the factors added for giving back the value for the money spent.

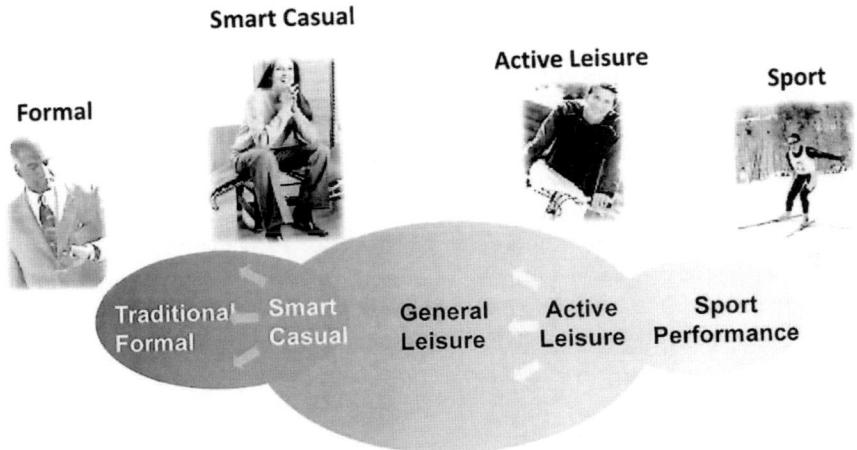

Figure 62: The trends

The apparel textile industry is undergoing through a paradigm change. The market today is more for casual dressings where pashmina will not fit, maybe some of the blends of pashmina or fine wool gets accommodated in this business scenario (Fig. 62).

For example, India is the youngest country in the world and obviously the market mechanism here is all for younger generation whereas in west Europe the trend is just opposite. There are a greater number of 1960's managers not so old and feeble in the marketplace and such senior citizens would like to use the garments with sporting look like Nike and Adidas t-shirts and the consumers feel satisfied as psychologically they feel more active and sportier. This is called passive sport.

8.1   China cashmere market report

In 2018, Chinese cashmere market produced about 15,000 tons and exported about 2000 tons in dehaired cashmere. China cashmere production accounts for about 60% of total cashmere produced in this world. The country has a cashmere goat population of about 120 million out of 700 million in the world. From January to June 2018 total export of dehaired cashmere was 1230 tons, an increase of 1.8% compared to the year 2017.

Total value is about 19 million USD which is about 17.8% more compared to 2017.

The price of cashmere in March 2019 was around 600 RMB per kg. The market kept quite since main buyers waiting to see market development. Towards the end of June and beginning of July 2018, prices picked by 850 RMB per kg. The reason for this increase in prices was that the Chinese government issued very strict environmental protection rules which led to decreasing number of breeding goats. The cost of feed increases at the same time quite heavily.

From July to September prices kept increasing slowly because it was a stable market without any big fluctuation. The business for cashmere yarn has been very positive both for exporting and domestic market. This situation supported the market price.

The production of cashmere and the goats in China slightly decreased based on lower cashmere goat numbers which lead to an increase of prices compared to 2017. Due to the changes in the environment the availability is low while fibre length is also shorter.

8.2   Mongolia cashmere market report

For Mongolia, cashmere is the world's third largest exporting industry after copper and gold. The cashmere industry provides the income to over hundred thousand peoples of whom 80% peoples are below the age of 35 and 95% of the peoples are women. Mongolia has an estimated total of 27 million goats and an animal cashmere production of 9400 tons.

The total production in 2018 was 5413 tons of clean cashmere (90% of total cashmere production) and 509 tons of dehaired cashmere.

The main importers of dehaired cashmere were Italy 409.5 tons, England 50.5 tons, China 38.4 tons, Germany 3.2 tons and Japan 1.3 tons. Majority of the production was first stage processing with final product manufacturing about 10%.

The government of Mongolia launched its new cashmere program in February 2018, the four years program focuses on domestic production of final cashmere products, technological innovation and financial support for cashmere produces.

Now that washing and dehairing process of cashmere production have been upgraded and the program focuses in spinning process, yarn production of knitted products and exports. The main Idea is to keep domestically grown cashmere and to keep final product within the country and reduce the wash cashmere in the china. From 2019 to 2020 the government plans to use 40% of raw cashmere in domestic production, improve present legislature of cashmere production and sales, increase in production and export volume through proper financial planning and support to business. The government plans to use 60% of total raw cashmere and improve productivity and quality of final cashmere with the use of advance modern technology. The cashmere combing industry will receive a working capital investment for 6–8 months while knitting will receive the investment for 2 years. The investment for technological innovation will be provided for 5 years.

8.3   Iran cashmere market report

In 2018, the Iranian cashmere markets have been severely hit through the devaluation of the Iranian local currency, gains some of its value back in October 2018. During this period, it was estimated that more than 60% of Iranian cashmere was bought by Chinese traders directly through there Iranian and Afghan agents.

Production for Iran remains unchanged. Quality also remain unchanged and production of worthy cashmere is estimated around thousand tons. In recent months, the prices of the meat and the demand of meat increased. The meat supply was not enough for the country, herders are then tempted to sell goats for the meat because the meat prices now are bigger than cashmere prices. These were taking a toll of quantity and prices.

Overview of major contributors to global pashmina

9.1 Pashmina from Mongolia

With the production of over 3000 tons of raw cashmere, Mongolia is the second largest producer of raw cashmere in the world after China. According to USAID, Mongolia is the second largest producer with about 15% of the world production as compared to China who takes over 75% and lesser percentages are shared by Iran, Afghanistan, US, Australia and much lower category is India and Nepal with about 1% of total production (Fig. 63).

Figure 63: Pashmina goats in Mongolia

9.1.1 History of pashmina in Mongolia

History of pashmina in Mongolia is not too old, it was during Mongolia socialist era (1921–1990) pashmina fibre was exported mainly to Europe and suddenly the whole business shifted dramatically to China in the democratic era since 1990 (Fig. 64).

Figure 64: Inspecting the pashmina fibre on goats back in Mongolia

This move from Europe to China benefited the People's Republic of China (PRC) manufacturing establishment enormously. With this tremendous growth of pashmina loans from Western and Japanese banking system, the Mongolian customers were unable to buy the Nomad's raw cashmere. Thus the nomads increasingly turned to Chinese.

9.1.2   Pashmina goat herders in Mongolia

Atypical Mongolia herder's owns about 1000 goats. A man with 400 goats is considered relatively wealthy. Cashmere trade have become particularly lucrative when price of kilogram of raw cashmere increased from 9USD to 40USD in the end of 1990. The harder's with the few dozen goat sell cashmere to middle men earning about 600 dollars per year (Fig. 65).

Figure 65: Pashmina goat herders in Mongolia

9.1.3   Economics of cashmere producing sector in Mongolia

According to USA about one third of population of Mongolia is engaged in herding pashmina goats as the part of their income. After the liberalisation of the herding section in the early 1990s to allow the private ownership of herds, the goat population increased dramatically, that is 5 million in 1990 to over 11 million in 1998. A price of 95 USD per kilogram is regarded as a breakeven point for a herder to make a profit by selling cashmere fibre. Prices of pashmina started rising in 2009 in parts because of harsh winters in Mongolia that resulted in animals frizzing to death. Herds were compelled to eat their goats rather keep them for undercoats (Fig. 66).

Figure 66: Modern Mongolian pashmina garments factory

The cashmere prices are always big news for Mongolia. Many herders use there radio to turn into information about the latest pricing. If the prices are good then they sell immediately and if prices are bad they store there produce in open air refrigerator and wait until they hope the price improve in goats (Fig. 67).

Figure 67: Economic activities for pashmina in Mongolia

9.1.4   Overgrazing by cashmere goats and desertification in Mongolia

Goats are regarded as harmful to the environment because they pull up and consume the roots of grass in addition to harming the environment, the loss of grass means the goats gets less from the land and quality of the pashmina decreases. Overgrazing leads to the desertification (Fig. 68).

Figure 68: Mongolian cashmere goats

According to UN development programs estimates that 90% of the Mongolia is fragile dry land under increasing threats from desertification. About 21% of Mongolia land mass is affected by increasing number of goat that leads to more desertification in Mongolia. About 80% of desertification in Mongolia is caused by humans and 30% by nature.

9.1.5   Cashmere industries in Mongolia

Cashmere industry and trade responsible for 15% of Mongolia's GDP. The garments and the value-added sectors are mostly underdeveloped (Fig. 69).

The objectives of the government of Mongolia for the cashmere industry are to develop cashmere industry so that it makes a significant contribution both to economic growth and poverty alleviation. Cashmere processing sector in Mongolia has seen significant export after the ban on exports were lifted in 1997. Another characteristic of the structure of the processing sector that influences the performance and future prospect of the sector specially on foreign ownership.

Figure 69: Pashmina production in Mongolia

9.1.6   Value added chain in the Mongolian cashmere industry

According to USAID, the value-added chain for cashmere in Mongolia has five main stages that are from raw material to processing to export marketing. As far back to 2004, the Mongolian export was about USD 57 million including USD 80 million minus imports of yarn worth USD 23 million. If smuggled quantity of Mongolian cashmere were included, then the net export would rise to about USD 97 million. Instead of smuggling if that quantity processed further and we go on adding value to a certain stage the Mongolian exporters could have earned more then what they are earning now (Fig. 70).

Figure 70: Pashmina marketing activities by Mongolian pashmina forum

The analysis raises the question that why smuggling is so rampant and why addition value is not been added, the answer lies in the cost of producing semi-finished and finished cashmere products in Mongolia. The Chinese labour cost is about 25% lower as compared to Mongolia. Looking into the details of value added chain for cashmere highlights the difficulty of the industry which exists now. Most cashmere is exported with only the low value addition and product quality is reflected in price. If this situation is not set right, the capital cost will be six times higher than normal (Fig. 71).

According to USAID, there are two lines of actions going on:

1. Herders,
2. Processing.

Figure 71: Pashmina goat herders

Both are in deep trouble. The herding sector may have surpassed the actual requirements of the herds. About more than half of the Mongolian raw cashmere is smuggled to China for processing in any way. Both segments of the market are highly distorted. The quality of the Mongolian raw cashmere is declined over time despite many projects to reverse this trend and even more recommendation in this trend reversing process. The average yield per goat declined. For example, reduction in 2microns in fibre diameter from 17.5 to 15.5microns which will raise the quality of pashmina from average to good but with the reduction in the yield which is by 24%.

The value of Mongolian cashmere apparel export increased to 9.6million dollars in 2016 and since 2009 there is not much of increase since in the global cashmere apparel industry. Desertification is not only threat to Mongolian herders but there is also a danger of global climate changes (Fig. 72).

Figure 72: Pashmina processing in Mongolia

In the recent past, Mongolia controlled nearly 30% of the global cashmere market and now that the market share is reduced and that of China within a period of 1990s and early 2000 the Chinese traders exploited falling wool prices to take over most of the Mongolian pashmina business.

9.2   Pashmina from China

In 2005 at the China International Cashmere Forum, it was made clear by the many speakers consisting of several government officials and industry representative that the Chinese outline their strategies for their cashmere industries. Worldwide dominance by controlling the supply and pricing of raw cashmere and being the "factory to the world" for finished cashmere products either as contract producers or with alliances with international brand holders and more importantly either by developing or buying brand names for themselves (Fig. 73).

Figure 73: Manual dehairing of pashmina in China

The report from USAID states that the government of China has subsidised its cashmere processing with very low interest rates loan maybe for last 40 years. Till today, no Chinese processor has been able to integrate forward into their international channels of distribution much less develop and internationally recognised brand name.

With the system of subsidising liberally the processing sector in China enjoyed for nearly 30 years at a stretch and that made them a powerhouse with which other countries are not able to compete. The Chinese government never ever tried to change the market for cashmere by using export taxes.

Figure 74: Pashmina goats in China

The pashmina/cashmere herding and processing sectors are largely located in Inner Mongolia, one of China's poorer region. As the Chinese government has granted substantial subsidise to the cashmere processing sector with low interest rate loans, tariff protection, no charges for workers welfare and housing and export subsidise in the form of tax rebates as a percentage of export sales in 1970s. There licencing system is very strict for domestic and foreign traders. When China joined the WTO (World Trade Organisation) in 2004–05, it was agreed to end the export incentive in 2004 and also ordered to reduce its subsidise on interest loans on bank loans (Fig. 74).

By 2005, China and Honking accounted for 80% of US knit imports up from 66% in 1995. China was keen to become the cashmere superpower in 1990s and then the cashmere fibre production went its fast driving for producing cashmere fibre from 9000 tons in 1990 to 12,000 tons in 1998 and they become super power.

In 1991, China with the total controlling power changing the market as they want, the country wanted to create confusion and problem when they withheld its entire supply from world market. In recent years the China has started banning cashmere goats from Inner Mongolia for environmental reasons, reducing the supply from 30% that is 7000 tons.

9.2.1   Cashmere goat herders in China

When the Chinese communist government dismantle the collective farming in 1994 the herders got back their grasslands. In Inner Mongolia, which is the poorest part of China, the most herders live in a permanent settlement and live as Mongolians in Mongolia. About 90% of the Mongolian herders are having the proud owner of colour television in their homes (Fig. 75).

Figure 75: Chinese pashmina goat herders

In this process of very high rewarding cashmere processing change the life of many herders. The affluence could be seen on the roads that everybody had gone much richer. A new middle class had come up who could wear pashmina garments and since then there was no look back for the herders. The herders grew prosperous. During the global financial crises, the price of cashmere fell almost by half and herders forced to sell many of their goats for the meat purpose.

9.2.2 Cashmere exports from China

According to Chinese custom statistics, China's cashmere and cashmere products export between January to July 2010 were valued at 552.7million USD in a situation where the growth of 25.2% year on year making the total including 1360tons of cashmere and dehaired wool was 95.42million USD, increasing by 69.1% and 78.6%, respectively year on year and included 457.28 worth of cashmere products increasing by 17.8% per year. A total of 2464tons worth 20.025million dollars of cashmere yarns were exported (Fig. 76).

Figure 76: Chinese pashmina products

The export of cashmere products realised the double-digit increase between January and July 2013. Raw material prices remain high but the selling prices for finished products do not adjust accordingly and the profits of the cashmere intensive processing enterprises do not increase with the rising export prices (Fig. 77).

Figure 77: Exports of pashmina goods out of China

The price rise of production factors such as raw materials leads to the rise of the cost of downstream products.

9.2.3   Cashmere industry in China

According to USAID, for decades the government of China has subsidised its cashmere processing with very low interest rate loans, reduced contributions to social benefits and until recently the export bounties. However, no Chinese processor has been able to integrate forward into the international channel of distribution much less develop and internationally recognised brand names (Fig. 78).

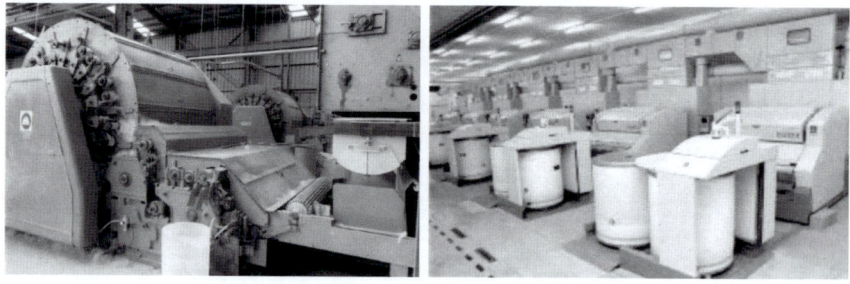

Figure 78: Pashmina industries in China

The processing sectors in China have enjoyed subsidising for over 30 years and has develop into a power house with which other countries is not able to compete.

9.2.4   Chinese processing sector

According to USAID, the Chinese labour cost (including fringe benefits) are about 25% higher in Mongolia while labour productivity is 25% less in China. China could develop the cashmere clusters of herders, processors, machinery and parts manufacturer, dye and other chemical input producers and traders and service producers and so on (Fig. 79).

Figure 79: Processing of pashmina in China

Overall at each stage of value-added chain, production cost is 30–40% higher in Mongolia as compared to China. The Chinese processors have on the other hand the advantage on the demand side. Chinese processors have several advantages over those in Mongolia. They have a large, protected domestic market where they can sell all the lower quality cashmere products.

9.2.5   Designers in China

The largest Chinese cashmere company is Eerduosi, it produces about one-third of the world cashmere. It is the largest cashmere producing company (Fig. 80).

Figure 80: Designing of pashmina in China

Cashmere sweaters sold in Britain includes Loro piana, Augustine, Uniqlo. Augustine Tse of cashmere house a designer known for his hooded sweaters, bath robes, baby clothes and even bikinis. Uniqlo V-neck and women prime mark jumper are few names which are in the marketplace now.

9.2.5   Desertification in China

The livestock population increased from 2 million in 1977 to 18 million in 2000 by turning one-third of the grassland area to desert. Desertification in China is primarily is due to over grazing and China needs to take some immediate measure to control the desertification seriously. The lands in China are too infertile to grow crops, herding is the only way for them to survive (Fig. 81).

Figure 81: Desertification in China

Since the 1960s, the average grass output has declined in inner Mongolia by one-third to two-third as the size of cashmere goat herds has increased. To prevent over grazing there is limit on number of animals that the livestock's herders can use. This disparity between the number of animals to feed and how much to feed, the supply and demand equation in this aspect is not clear.

The way UAE is trying to make their country clean, green and beautiful; it is a subject matter which needs to be ponder about. The same technology can be used probable to make the soil suitable for growing fruits and vegetables in China as well.

Causes

The main causes for the desertification in China are over grazing, over planting, over overflowing and raising crops in the region which is too dry for the crops. Some thoughts portray the main cause of desertification is the then the leader of the country Mao Tse Tung planning to grow grains in the area where grains are not growing well.

Some school of thoughts believes that desertification is caused by climate changes. Scientists from Qunghai have recorded much higher temperature lower rainfall and stronger winds since 1950. Persistent drought robs the soil of moisture and makes it easier for the soil to pick up and carry away by the wind. In Inner Mongolia, global warming appears to have made the region dry.

9.3   Pashmina from Afghanistan

9.3.1 Cashmere production in Afghanistan

The Afghan cashmere industry is estimated to be about 18million USD in export of greasy cashmere. In most of the stages along with the value chain are conducted outside of Afghanistan. Afghanistan itself is involved only in production and harvesting of pashmina fibre (Fig. 82).

Figure 82: Pashmina goats of Afghanistan

Only exception is an Afghan-Chinese joint venture in the city of Herat, which has established a scouring line and planned to install dehairing equipment's adequately (Fig. 83).

There are two types of Afghan greasy cashmere at raw stage:

1. Bahari cashmere which is also known as Spring cashmere. This fibre is relatively clean with no chemicals applied on it. It is short in fibre length and brittle. The value of Bahari cashmere is substantially higher than of skin cashmere.

Figure 83: Processing of Afghan pashmina fibre

The Herat exporters tend to fix these qualities together, in a certain ratio often in a 30:70 or 40:60 to reduce the price of the overall products. Some consignment is also kept as 100% Bahari cashmere.

2. Skin cashmere – this is a tannery cashmere obtained from the dead skin of cashmere goats. Chemicals applied to the skin, uncleansed, also shorter in fibre length and more brittle. This type of cashmere is produced from winters to early spring period.

Export volume from Herat which only real trading hub for cashmere is, which has reached up to 930metric tons in 2006 and 2007.

9.3.2   Methods of harvesting

In Afghanistan there are two methods of harvesting:

(a) Combing

(b) Shearing and hand dehairing.

Shearing is by far the most prevalent method of harvesting in Afghanistan. The goat hair maybe sold or spun into yarn for making ropes, tents, rugs or even for the lining of carpets due to its insect repealing characteristics. Combing is an acceptable or preferable process by a small number of farmers notable in Badghis province and to less Kandahar, Nimoroz, Zabul and Uruzgan. The international experience describes the advantages of combing and strongly recommended that the Afghan farmers change their harvesting techniques to combing. Combing cashmere is cleaner, has high yield ratio and has a longer fibre length. The longer fibre length stems from the fact that the cashmere often gets cut during shearing which reduces their fibre length (Fig. 84).

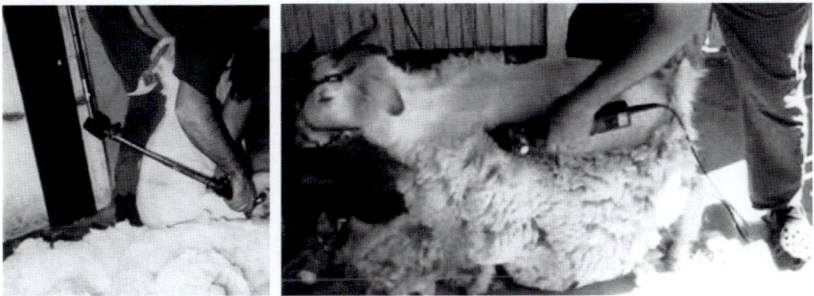

Figure 84: Shearing of pashmina goat of Afghanistan

However, the Afghan exporters do not see the advantages of combing with some claims that combing will reduce the quality of cashmere because it will cause the fibre to break. There is, therefore, no price premium paid for the combed cashmere in Afghanistan. Some exporters may even pay low cost for combed cashmere (Fig. 85).

Figure 85: Combing operation of pashmina goats

Some experiment was done to establish a contentious opinion from the difference in views about the same.

To settle the difference in views as soon as possible conducted two experiment to compare the two-harvesting method. First small experiment considered on comparing yield and time required per goat when harvested or combed. Each goat was shorn on one side and combed on the other side and the weight of the cashmere plus the time required for combing and shearing was compared in addition the men and women were interviewed regarding their views and preferences. The second experiment aimed at obtaining more scientific data on the effect of combing and shearing on yield, fibre length and diameter in dehaired cashmere.

9.4   Pashmina from India

Pashmina fibre is considered as a creation of India. The goats (Changthangi, "Changra") are very typical parts of trans-Himalayan mountain region where the temperature is average −42°C. Pashmina from India have been valued for centuries throughout the Asia and middle east and the wonderful qualities of pashmina is also popularly known as cashmere fibre. The cashmere and pashmina are same fibre with same chemical structure. As a creator of this fibre, India does not fit sizable in the world pashmina/cashmere production from the very beginning (Fig. 86).

Figure 86: Pashmina from India

While the world production is somewhere about 20 thousand tones, India has a share barely 0.28% even Nepal produces pashmina fibre slightly more than India. 72% is from China, 18% from Mongolia, 7% from Afghanistan. They are the major players in the pashmina trade and industry.

Afghanistan and Inner Mongolia beside China of course are now quite ahead in the growth path. The government from all three countries including India and Nepal are all started giving importance to some extent to the development of this industry which produces a wonderful fibre pashmina – natures wonder.

9.4.1   History and economics of Indian pashmina

In India, pashmina is obtained from Ladakh region of Jammu and Kashmir, Lahul and Spiti valley of Himachal Pradesh and some parts of Uttaranchal. Indian Pashmina shawls and knit wear are always having a good demand since the ages of Mahabharata (Fig. 87).

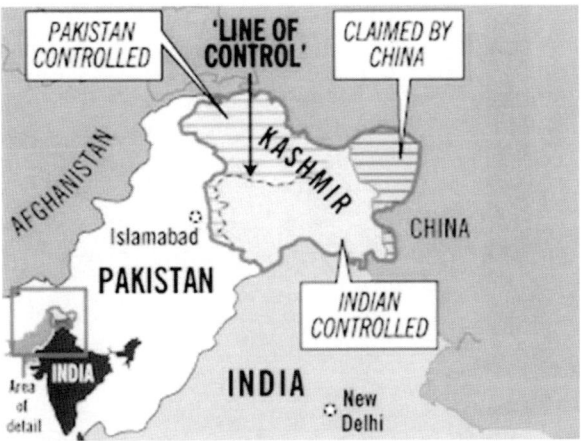

Figure 87: Map of the pashmina activity centres

The Indian hand spinners and weavers were the best in the world. The real start of getting attention of the international modern consumers was in the beginning of 1990s. Precisely, in 1995 the pashmina shawls production was absolutely at the peak and for the next 10 long years, India did extremely well in the world market (Fig. 88).

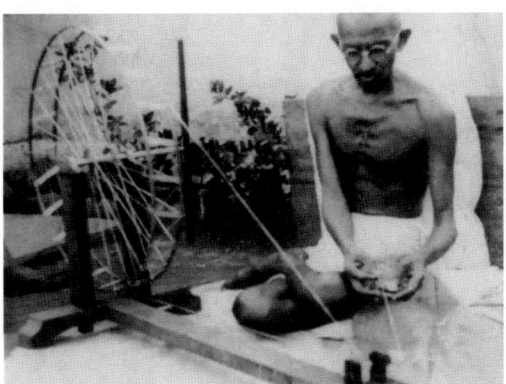

Figure 88: Mahatma Gandhi ji on pashmina hand spinning

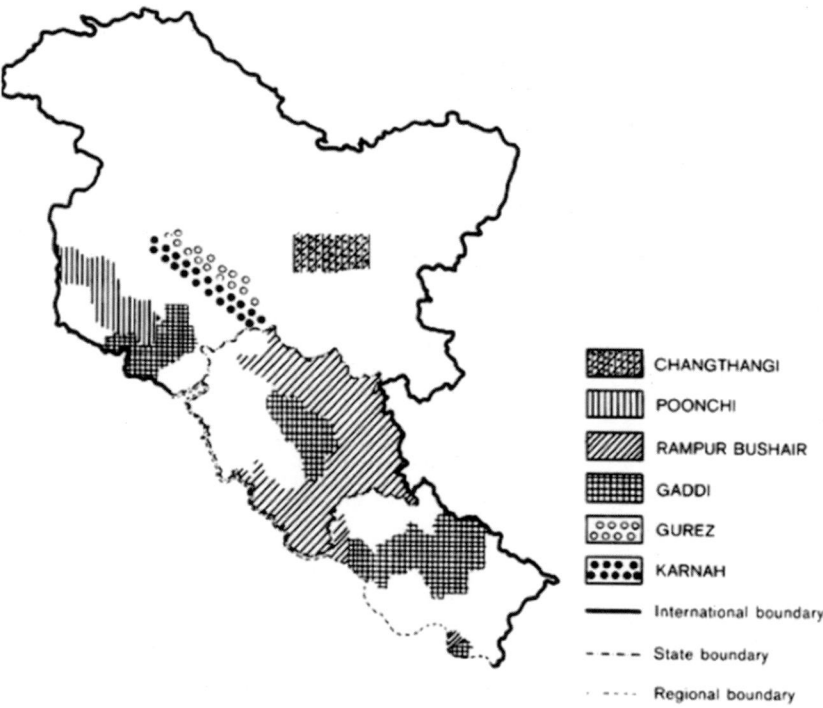

CHANGTHANGI

POONCHI

RAMPUR BUSHAIR

GADDI

GUREZ

KARNAH

International boundary

State boundary

Regional boundary

Figure 89: Sheep and goat breeds of India

This 10 year could change the business mind of the shawl manufacturer and they have become real international business partners with modern amenities, with due attention given to them much needed innovation. As a matter of fact, those 10 years business experience made them completely different from their previous situation. Cashmere more precisely in Ladakh region grow about 45 tons of pashmina fibre every year from the population of about 2.50 lakh of pashmina goats (*Capra hircus laniger*) (Fig. 89).

The local industries do not consume all of it as nearly a part of the total pashmina fibre production gets sold out of the country.

Shawls as such are always used by the middle-aged women of India, Bangladesh, Pakistan, Middle Eastern countries, etc. Pashmina Shawl is a product which were meant for the upper-class consumers only because of its high price (Fig. 90).

Figure 90: Retailing of pashmina shawls

Only in 1990s, the fate of pashmina shawl industry got a boost after introducing pashmina shawls as an international fashion accessory by modifying the existing shawls to a stole of European design and colour. The 10 years from 1995 to 2005, the shawls industry worldwide made a paradigm shift in the total shawl business. But 2005 onwards they started seeing slow decline in this business. This maybe because of the market mechanism of that period and also by the phenomenon of the fashion cycle. Any product however, good they must have to come down suddenly after a period of extremely good business.

The pashmina industry in India is considered too small at the present time to attract the attention of government of India. However, the government of India in there 12th five years plan in which the Pashmina is considered as an important and potential area of textile to work on. To attain this level the main reason is the wonderful quality of the fibre with excellent feel and handle.

India is a shade better in their processing with the sophisticated machineries and innovative marketing than its neighbours.

9.4.2   Growing of pashmina fibre in India

Ladakh as mentioned earlier produces one of the better pashmina fabric in the world but the production is very tiny and limited. There remains a dispute of exact amount of pashmina production in Ladakh and other region which means with the total production is unconfirmed. But the figure of 40–50tonnes per year of Indian pashmina with 2,00,000 pashmina goats are available. The net value of dehaired pashmina is approximately 40crores per year (1cr = 214431.36USD) (Fig. 91).

Figure 91: Herding of pashmina

In India, combing is the major method used for harvesting pashmina by using a special type of comb. After harvesting, the pashmina fibre is dusted manually to remove all the impurities like sand, dust, etc., which accounts for nearly 10–20% of the raw fibre weight.

9.4.3   Sorting and grading of pashmina fibre in India

Sorting is a process by which the mixed-up contaminants are separated out. This process is done by village women's who are very well trained for their job. The quality of shorting is very much depending on the individual and their capabilities (Fig. 92).

Figure 92: Sorting and grading of pashmina shawls

Grading is a process very commonly used in wool industry and this is a very important component in processing merino wool which comes from Australia and lot of emphasis is given on its usefulness. Grading is one of those subjective quality upgradations which is vital for the ultimate quality of the product. This process cannot be ignored, or half done as the ultimate quality of the shawl which will make differences if the sorting and grading are not done properly.

India needs some attention at this stage of processing.

9.5   Pashmina from Nepal

Nepal is another promising pashmina business centre coming up very soon. Quite a few important projects with United nations fund was successfully done and the outcome of all the project put together tells a story that Nepal must go a long way to achieve the desired result, but they are confident enough to turn around. Like India, Nepal also has the Himalayan Cashmere goats from which they produce about 40–45million kilograms of pashmina. They have 200 thousand goats and the output is stated above. This is abnormally low compared to the global standards. Nepal does not have any sophisticated machineries; they do not have proper test house and most importantly they do not have an upgraded international standard Fashion institute. These are the three important criteria to increase the production of pashmina keeping the quality standard (Fig. 93).

Figure 93: Nepal pashmina industry

The natural colour of the fibre is white, grey and brown. The white colour pashmina are most preferable but expensive. The average pashmina production of Changra fibre is somewhere like 200g per goat per year with average production (Fig. 94).

Figure 94: Pashmina shawls of Nepal

To illustrate, global market prices for washed and de-haired cashmere fibres in July 2007 is:

(a) Chinese white sells for US $ 100 per kilo,

(b) Tibetan brown sells for US $ 97 per kilo,

(c) Mongolian brown sells for US $ 85 per kilo.

All prices are to be updated but just for information set and showing the trend, the above figures are considered.

9.5.1   Economics of pashmina in Nepal

The domestic market of Nepal is very limited for pashmina to the extent of the number of tourists coming to Nepal. The pashmina market in Kathmandu is made for the foreign tourists visiting Nepal. The big earthquake happened a couple of years back and put a hold to the growth of the domestic market quite considerably. The US market however is a major importer for the cashmere particularly for the medium and the lower end segments. In 2006, US imported over 13 million cashmere sweaters for a value of 494million USD (Fig. 95).

Figure 95: Domestic market for pashmina product

The export of Changra cashmere shawls boomed in the late nineties as mentioned earlier, recording a figure of 82 million USD in 2001 and then suddenly spiralled downward in 2005/06/07. The export came down to below one fourth of the peak record of the early thousands. The main reason told by the manufacturer and exporters of pashmina shawls is the normal cycle of fashion items. Any fashion item for that matter has a life cycle that it comes up quickly and after a decade of business it crushes down. This is what happens primarily. The unhealthy competition from the neighbouring countries and which used much lower quality of raw materials and making a fake pashmina shawl which is normally sold at very lower prices than normal Nepal market price (Fig. 96).

Figure 96: Changra goat for pashmina in Nepal

The world demand for Changra fibre is much bigger than the world could supply. A concerted effort of a stake holder aimed at increasing the Changra cashmere product export through a program of improving production, sourcing and testing the quality of raw materials. The introduction of innovative new items and registering "Changra cashmere" as recognised or certified trademark. The popular blends of Changra cashmere and silk has seen very good days in the export market in 2006–2007. These include shawls, stoles, knit wear garments and accessories.

The real cashmere fibres of Nepal are found in white, grey and brown colours. The diameter of these fibres is 11–19 microns, and length is around 25–50 mm. According to US regulation the animal fibres higher than 19microns and shorter than 25mm can no longer entre into the country as cashmere. The accepted international standard for cashmere includes anything less than 19microns thick (Fig. 97).

Figure 97: Not really 100% pashmina but blends

The word pashmina is being used all over the world for shawls with soft acrylic in the blends. This has put Nepal and India as the rogue markets. This has created the downfall of pashmina shawls and stole. To build the image back again, it takes a long time and huge amount of promotional activities and money spend for convincing the peoples to come back to the glorious name and fame of the pashmina fibre back. Nepal and India both are very conscious and started putting efforts accordingly.

9.5.2   Shortcomings that Nepal must address

Having said all about Nepal pashmina business, it is necessary to highlight the shortcomings of the country and its pashmina business (Fig. 98).

Figure 98: The signature of Nepal pashmina products

1. Image building
    Image building activities like attending all the important exhibition which are:
    *   The maison and object trade show.
    *   The cashmere world.
    *   The Heimtexil.
    *   The millionaire fairs.
    *   The 100% Design Trade Fair.
2. Nepal should have its own signature.
3. Nepal should have a modern spinning unit to be build.
4. Nepal should create its own proper modern finishing lines.
5. The Country should have an international standard fashion and design institutes.
6. The country should have a justified and credible test house.
7. The country should have a mechanism of creating trained and skill workers on a continuous basis.
8. The pashmina boards or forums or agencies (government as well as private) should have their eyes open as per as the innovation in each and every product of pashmina and to the eco friendliness in the whole supply chain.

9.6   Pashmina from Australia

Pashmina is the finest and softest of all textile fibre. It is substantially lighter and warmer than wool of the same fineness. Pashmina therefore occupies a top position and the price reflects that position. Traditionally, the cashmere fibre produced by China, Mongolia, Afghanistan. The qualities of cashmere from these countries are poor because of the contamination with dirt and coloury fibres (Fig. 99).

Figure 99: Australian pashmina breed

Australia/New Zealand animal husbandry practices and the open spaces there allow them to produce a well-nourished fibre that has a unique life and feel. The worsted yarn produced using this fibre has a strength and vitality never seen before.

Figure 100: Australian white cashmere

Australian cashmere is a unique, sensuous and rare natural fibre. Australia is relatively a new entrant in this trade but doing extremely good as far as the pashmina fibre production is concerned. The amazing discovery in 1973 among the feral goats of western New South Wales of animals bearing a surprisingly good coverage of cashmere beneath their other uninteresting exterior coats of coarse hair (Fig. 100).

The cashmere is valued for its production in small quantity for downing undergrowth called "Duvet" by the French, this grows with a coarse long guard hair. The duvet is extremely valuable. The duvet is sheared in the spring and is carefully combed out. These goats are known as the goat of Tibet or of the valleys of Kashmir. Small flock of pure cashmere descended from an importation of a buck and two does was brought from India by an agent of the Peninsula oriental company in 1873. He ran them very successful in longer run at Longerong in the Wimmera of western Victoria.

In 1874, cashmere fibre is taken to Tasmania. The fleece bearing goats were reintroduced to south Wales in 1897 from animals of Tasmania which were descended of the importation by a society of Victoria in 1870 (Fig. 101).

Figure 101: Separating pure cashmere from courser hair

Pure cashmere for commercial purpose is obtained by separating the down from the courser hair. The diameter of the Asiatic Cashmere average from 14.3 to 22 microns and the length from 3.2 to 6.1cm with which the newly discord down compared favourable.

In 1973, CSIRO researcher did find that feral goats of New South Wales carried cashmere fibre. After further investigation it was estimated that 20% of feral goat herd can produce useful quantity of cashmere. Recently a suggestion came from west Australia that at least 30% of feral goat produce reasonable quantity of cashmere (Fig. 102).

Figure 102: Feral goat of Tasmania

There are an estimated 2 million feral goats running throughout Australia. Western Australia has the nation highest feral goat population and it is only state to declare feral goats as vermin.

Cashmere is the most expensive fibre in the garment industry, with processors paying 10USD for quality white cashmere. A cashmere fibre from Australia has a course wavy structure with no cream. It is tough, machine washable and non-felting.

9.7   Pashmina from Iran

Iran is another relatively small but advance pashmina herding country. Annual production of cashmere of Iran is 0.01% of the world Textile Market. Cashmere production and harvesting is difficult and labour intensive as the quantity produced is very limited, the price of this luxury fibre is very high. Being expensive, cashmere fibre necessarily have a market which is limited to only to rich consumers who want the best of the world material.

Iran has about 25 million goats, 5 million are cashmere producing while remaining goats produce small amount of cashmere. It is estimated that the Iran cashmere goats produce about 2000 tons of cashmere annually. This quantity of cashmere is either exported as raw undehaired (70%) or processed (30%).

40% of all goats are kept by nomads in habitat about 95% of the total area of the country. More than 30% of Iranian cashmere is produced by Raeini and Birjandi goats in Kerman and South Khorasan Razavi provinces, respectively. The quality of Iranian cashmere being long and crimpy which quite a good spinnable fibre is. Cashmere in Iran is the down fibre driven from the fleece of a domesticated goat (*Capra hircus*) descended from genes *Capra aegargus*, the wild goat of Persia. In China, Mongolia and parts of Iran cashmere is harvested by combing during the 3–6 weeks spring period when the goats are moulting or collecting the moulted fibre from the ground and the bushes (Fig. 103).

Some study indicated that the average fleece weight per shearing per goat is 540 g in a range of 100–700 g. 45% of the goats produce 550–700 g of cashmere while 7% of remaining goats produce 100–250 g.

There are many advantages of combing and it is strongly recommended by experts that farmers change their harvesting technique from shearing to combing. Combed cashmere is cleaner with higher yield and longer fibre length, arises from the fact that 1–2 mm of cashmere is left on the body of goats during shearing.

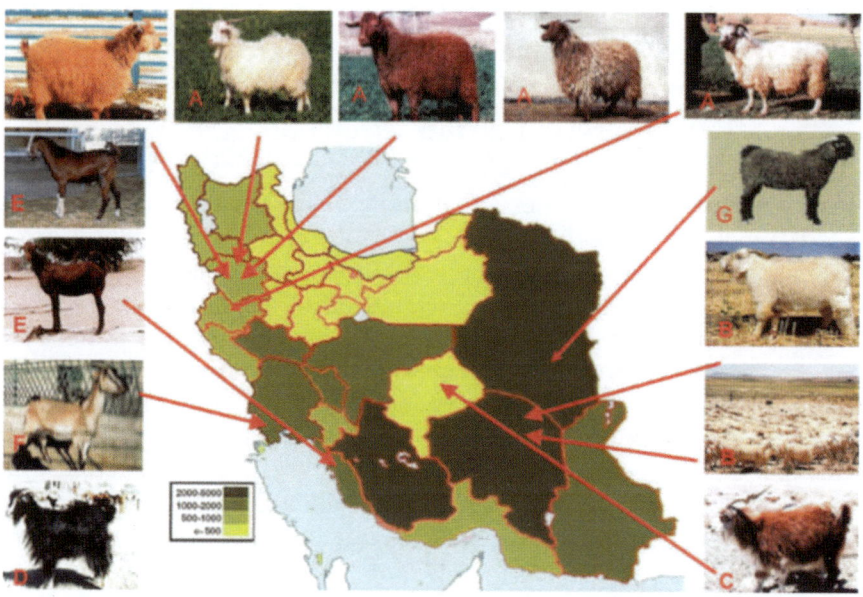

Figure 103: Goat breeds distribution by region in Iran (A, Marghoz; B, Raeini; C, Nadushan; D, Hairy black; E, Tali; F, Najdi and G, Birjandi)

9.7.1   Marketing of cashmere

The world cashmere clip is marketed by direct buying or speculative buying, either cooperative marketing through private cells or sealed bid cells which auction is selling by open comparative bidding.

In Iran, practically nearly all the cashmere is marketed by direct buying by processors and manufacturers. After harvesting, cashmere is brought directly by the herders by middlemen and it is directly stored in warehouses according to colour and fineness. The most important characteristics while buying cashmere is still the diameter. The Iranian government time to time take actions to encourage the export of processed cashmere with considerable value addition. Being a truly luxury fibre, the demand is greatly affected by the fashion changes. International processers pay according to the quality which is primarily determined by the fineness (fibre diameter) measured in microns. Some indicative prices are shown below (Table 15, Fig. 104).

Table 15: Prices paid for Iranian Cashmere

Micron diameter	Colour	USD kg
15	White	50–140
15	Coloured	45–130
16	White	50–130
16	Coloured	45–120
17	White	40–115
17	Coloured	35–100

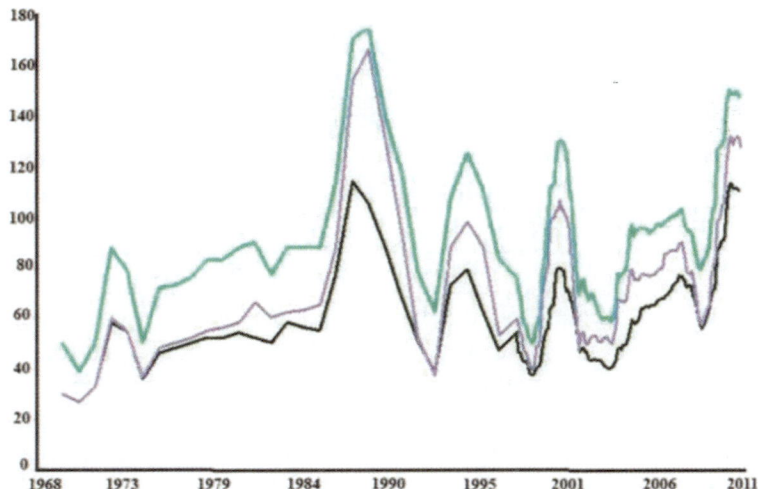

Figure 104: Cashmere price of China, Mongolia and Iran over time. Chinese cashmere is sold at a higher price than Mongolian (second highest) and Iranian cashmere (Schneider, 2012)

The overriding influence on the price is the mean fibre diameter. Colour is also an important factor. Value added to cashmere in the several stages of processing will make the final price for clean, value added, dehaired and spun cashmere can be up to four times the raw greasy price received by the herders. The Iranian cashmere is generally 2–3 microns coarser than Chinese and Mongolian cashmere and longer and crimper then Australian cashmere. This makes Iranian cashmere cheaper.

The average Iran cashmere staple fibre length is 54.2 mm and the diameter are 19.7 micron ± 1.5 micron and these specifications will be good enough to be worthy for worsted and semi-worsted industries and therefore, the Raceini goats would qualify for worsted and semi-worsted industries.

9.7.2   Cashmere value addition

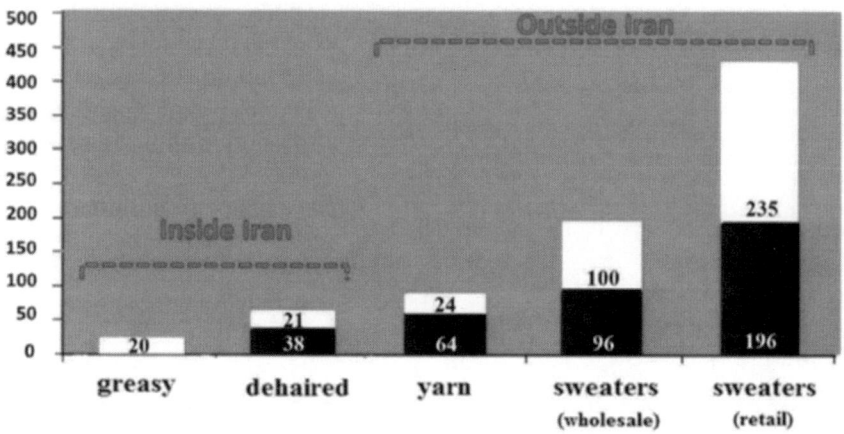

Figure 105: Value added chain

In the Iran cashmere value chain, the value addition for each step in the processing is as shown in the fibre 97. At the lowest section of the value-added chain considerable potential exist to expand production of cashmere to goat farms in other provinces. Raceini goat which is the main cashmere producing breed of Iran can be characterised by having long and highly curbed cashmere. The steps must be taken to improve the fibre diameter to capture the higher prices in international market (Fig. 105).

9.7.3   Processing of pashmina in Iran

The Iranian cashmere fibre is also keratin fibre. The chemicals and physical structures are very similar to that of wool. Hence, the cashmere fibres in Iran are dyed in the similar sort of dye stuff the wool processed by. The trading of cashmere product is on the north east region feeding into the main centre of Herat near Iranian border. Sorters employed by the Herat traders manually dehaired cashmere and then exported to Belgium. Major processing centres of Iranian cashmere are Semnan and Mashhad cities. Belgium has the only large-scale European facilities for disinfecting raw animal fibres which comes from Iran, Afghanistan, particularly in the case of cashmere emerging from Afghanistan and Iran (Fig. 106).

There is always a risk of infecting with Anthrax. The Chinese buyers appear in the Herat markets and as a result the price for cashmere increased by 10% over year. The processing part is being carried off in Kazakhstan, where in 2005 the first large scale dehairing factory oven in city of Chimkent and processed many tons of raw cashmere send by the trains of Afghanistan. Normally the Chinese machineries are used in dehairing and they get a reliable yield of nearly 65% of shorn goat fibre. The raw cashmere in Iran is processed in several stages – hand dehairing, scouring, machine dehairing, spinning, weaving and knitting.

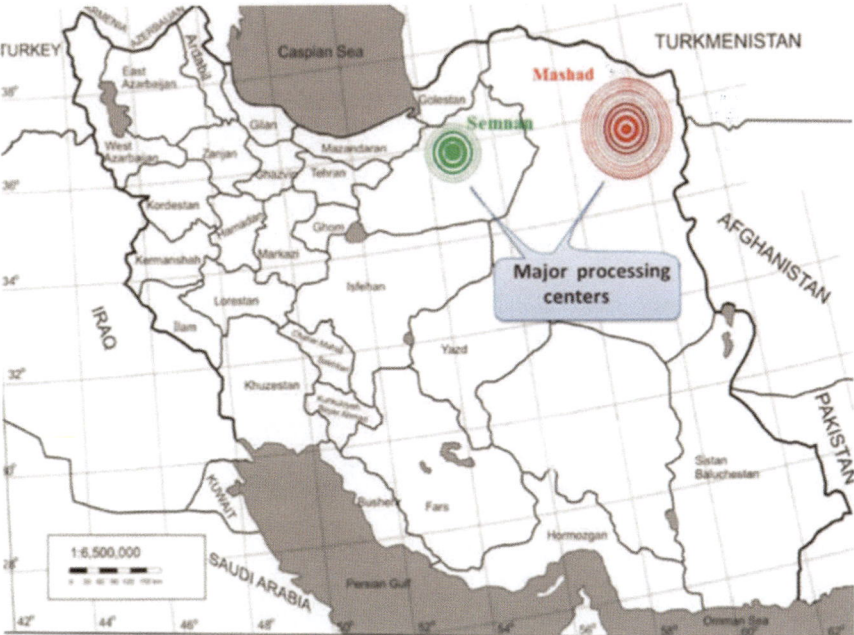

Figure 106: Major processing centres of Iranian cashmere

The raw pashmina fibre reaches the mills in the form of fleeces which are packet and transported in jute or polymer sacs. The weight of the bale varies greatly and often this indicates the origin of the cashmere fibre. The cashmere fibre from Kemal is loosely packed in rectangular jute sacs or tall narrow polyesters bags weighing about 100–150 kg (Table 16).

Table 16: Stages of cashmere processing

Stage of processing	Place of processing	Personal involved	Society level
Cashmere combed or shorn from the goat	Country or region of origin	Local farmers or travelling cashmere buyers	Village farms
Cashmere is sorted by hand into colour and/or diameter	Country or region of origin or neighbouring country in Central Asia	Local women and men	Villages town or city
Cashmere is dehaired by hand	Country or region of origin of other country	Factory workers (usually male)	Larger town or cities
Cashmere is willowed to remove dust	Country or region of origin or other country	Factory workers (usually male)	Larger town or cities
Cashmere is scoured (washed in hot water) to remove dust, grease, dung	Factory in country or region of origin or other country	Factory workers (usually male)	Larger town or cities
Cashmere is dehaired by machine	Factory in country or region of origin or other country	Factory workers (usually male)	Larger town or cities
Cashmere is carded for woollen spinning	Factory in country or region of origin or other country	Factory workers (usually male)	Larger town or cities
Cashmere is made into tops for worsted spinning	Country or region of origin or other country	Factory workers (usually male)	Larger town or cities
Cashmere is spun into yarn	Country or region of origin or other country	Factory workers (usually male)	Larger town or cities
Cashmere is woven or knitted into garments	Country or region of origin or other country	Factory workers (usually male)	Larger town or cities

Before scoring individual cashmere, bales are often sorted for style, quality and some other contaminants which must be removed at this stage which has a direct impact on the qualities of yarn produced. The sorting is an important process as it determines the quality of final yarn which should be contaminants (sand, dirt, burrs, pollen and many other vegetable impurities) free. The cashmere fibre is processed through a double cylinder opener for dusting or opening an especially dirty, long and course cashmere (Fig. 107).

Figure 107: Bales of different colours of cashmere transported from production sites and kept in warehouse at Mashad processing mill

The scouring is an important process in cashmere production which removes again the vegetable impurities, the suint, dirt and dung attached to the fibre or entangled with the fibre and the residual grease content. Normally, a modern scouring process will use non-ionic detergent than only alkali and soap as the cleaning detergents (Fig. 108).

Figure 108: Cashmere sorting area

The dying of the washed cashmere fibre then dried through a dryer which runs on the flow of hot air produced by steam pipes placed either in the separate compartment or in the same dryer itself. Normally the residual moisture content of the cashmere fibre immediately after coming out from the oven should be somewhere between 9% and 10%.

The other steps are pre carding processing like garneting and burr-picking, carding and dehairing. The materials are then packed in bale formed to be exported to Europe.

Cashmere is often considered as a difficult fibre to process because of the presence of guard hair and cashmere in the fleece together (Fig. 109).

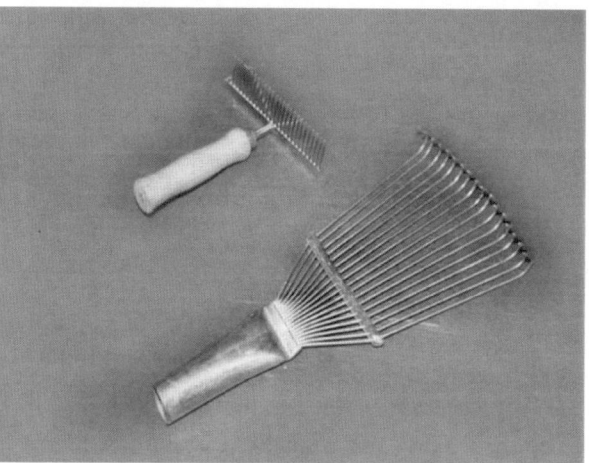

Figure 109: Short and long comb

To control the quality of the whole process some in-house testing are necessary so a capable person and a small laboratory is a must for all the manufacturers. In the quality check they normally check the residual grease content on the scoured fibre, moisture content in the dried fibre after drying, guard hair percentage, fineness and staple length.

Some definitions which are commonly used in this trade

- Pashmina and cashmere are the same term used interchangeable according to the use of the region.

- Raw pashmina/cashmere fibre which are uncleaned pashmina/cashmere fibre having guard hair as it is taken from the undercoat of goat.

- Scoured pashmina/cashmere fibre: Pashmina/cashmere fibre washed, cleaned by mechanical and/or chemical method which removes impurities and foreign matters.

- Sliver/top: An indefinitely long assembly of staple fibres, substantially parallel and without twist and capable of being drafted.

- Pashmina (cashmere) yarn: A yarn made of pashmina fibre, spun on the woollen as well as worsted system.

- Woven fabric: A fabric produced by interlacing (by weaving on a loom or a weaving machine) a set of warp threads and a set of wrap threads and a set of weft threads and a set of weft threads normally at right angles to each other.

- Knitted fabric: Fabric constructed by interlocking a series of loops of one or more yarns by hand or by machine.

- Needle/punch felt: Piercing tufts of raw wool hundreds of times using a very sharp needle with tiny barb to sculpt varies shapes and figures.

The most current issues and their implications on the pashmina trade

The main issues in major pashmina producing parts of the world are as follows:

1. Herding

   (a) Goats and herders living conditions are very pathetic which needs immediate attention. There are no proper or adequate housing facilities for herders and for the goats.

   (b) No suitable water supply for herders and the goats. This impact on the quality of the pashmina fibre.

   (c) No necessary medical facilities up there.

2. Some countries have all the sophisticated machineries required but most of the cashmere shawl production is handmade. The machineries and equipment need to be modernised as in the global market it is impossible for cottage industries to survive.

3. Cashmere shawls and shawls from other countries are all in the need of innovation at every stage of their production. Innovation is the main criteria in the business today.

   Innovation is not enough, and it needs to be the top priority.

4. All the pashmina producing countries should have their own independent training, education, skill development centres.

5. All the pashmina producing countries should have their different signatures on their products to identify the sources.

6. Tie up of industries and academia on interactive bases.

7. A real image building activity especially for the pashmina manufacturers. We need to change our image the very best in the world. Somehow, the image of the pashmina product which is very expansive material and very wonderful material should be built slowly but steadily.

8. The data on pashmina is not available freely. It is almost impossible to get updated data on cashmere or pashmina that's why how we keep our business going without any data, it is like someone hitting a stone dark in the night. Hence, there should be a data bank on pashmina products and marketing, necessarily to be made to support the growth of pashmina sectors globally.

9. Most common issue faced in shawl producing countries like China, Mongolia, Iran, India and Afghanistan is low availability of literature on pashmina.

10. When France, the biggest buyer of pashmina lost the 1870–71 Franco-Prussian war, the trade in cashmere service came to a halt.

11. All the above issues need to be addressed too. It is believe that for every problem there is a solution to it.

12. There are enormous number of issues on the pashmina production and trading related to the government laws and regulations, export marketing activities, domestic issues related to the pashmina trade, issues on climatic conditions, export barriers, etc.

Intellectual property rights of pashmina trade and industry

12.1 Economic impacts of GI registration of unique textiles in India

The geographical indicator (GI) is now quite common to the members of the intellectual property rights under Trips agreement signed by WTO and its members in 1994. The agreement defines GI as an indication that identifies goods originating in the territory of member or region or a locality. The given quality, reputation or other relevant characteristics of the product like pashmina is essentially attributable to its geographical origin. Hence, GI as defined in the agreement emphasis few attributes like qualities, characteristics, the belonging of the product, etc.

The economic rational of IPR protection through GI is closely associated with the perception to link origin as quality signal for marketing the goods. It ensures the origin and unique quality of the product for development. It also protects the consumers from deception by eliminating quality – prize disparity and infringed goods. The registration of the product under the GI act provides an important legal tool to the properties, to initiate infringement the actions against the countries legal system.

12.1.1 TRIPs agreement and India's GI act

"The TRIP" agreement while defining GI gives emphasis to quality, reputation or characteristics to the goods. It was intended to link to the geographical origin for protection. It means when some geographical regions acquire reputation for origin of products for unique quality, it is the quality of the reputation that distinguish product from other products.

Further, when a product acquires such reputation there may be an attempt by others to indulge in unfair business practices to capitalise the premium prize associated with it.

General level production (Article-22) is given to member countries to legitimate specific rules for absolute protect to wine and sprit and general protection other than wine and spirit. Such a dual protection mechanism has been questioned by many other countries, even if the member countries accepted the TRIP's mandate on absolute protection, majority of the countries have been insisting for same level of protection for their products. It was initialled in year 2000 when the country like Bangladesh, Cuba, Sypris, Pakistan, India, Nepal and others opposed the descriptive nature of additional production and demanded similar protection for the other products. However, their agreement was neglected on the basis of Article 24.1 which stimulates negotiation for enhanced protection for wines and spirit only.

However, the popularity and market penetration for the product increased several times as soon as the Pochampally Ikat got registered under GI act 1999. The IPR protection brought the visibly to the producers and the product. The product got noticed because of the GI registered company both in domestic and international market due to publicity by mass media and its popularity.

The constitution of an inspection mechanism for monitoring the quality and other key variable as stipulated in the geographical indication (GI).

This whole process has also helped the weavers in standardising their products across the production chain. Earlier, the weavers were not ready to share the information with fellow weavers due to unhealthy competition among the weavers. Since, the GI has adopted it shows a better result in craftsmanship and any innovation and diversification of the product for better consumer satisfaction which in turn create more demand for both domestic and international market.

12.2   Wool mark

Woolmark is a quality certification trademark. It was earlier owned by international wool secretariat and now under the Australian Wool Innovation. Though Woolmark is meant for 100% Pure wool, they also consider other animal fibres worthy for having the same Mark, if the properties laid down by the company confirms to their Standards. (Fig 110)

WOOLMARK

Figure 110: Certification Trade Mark

Normally and generally Woolmark is for 100% pure new wool. It is believed from the information sources that 100% animal hair like pashmina and cashmere fibre can be considered for being Woolmark licensee.

The IWTO (International Wool Textile Organisation) report is an authentic test procedure to be adopted for Woolmark licensee.

This multinational organisation with their marketing tool 'Woolmark' which is a certification trademark is very well-known mark, this mark has a tremendous brand value. Having the Woolmark licence is like joining an International club which will give the licensee a strong back to enter successfully in the international market. There are several marks beside "Woolmark" like wool blend marks – this mark is given to companies who are using minimum of 30% wool and rest maybe synthetic fibres. The blend ratio of 55:45, 55% wool and 45% other fibres is considered as the best business ratio in the organised sector for suiting fabric making industries.

Government intervention in pashmina industry and its implications

13.1 Mongolia

In 1917 when government of Mongolia began a campaign to increase the quantity of cashmere and meat production also

This order came from government of Mongolia. Put quite a heavy impact on the quality and prices eventually, it was to satisfy the consumers demand. There was a tremendous pressure on the herders to increase the production immediately.

Afghan was the country devastated and could not maintain the supply of pashmina for nearly two and the half years.

A new breed of goats has been created by Afghan Mongolia 'Gobhi goats' which has large very.

The Russian don goats are also related to pashmina and angora goats producing Angora (cross beard) which ultimately produce a product Angora.

Cross product has courser fibre than the Mongolian pure breeds. By then the Mongolia achieved independence and privatised its herds. Much of the country retained significant population of native Golan goats.

The export ban in 1994 created the market distortion and encouraged it regarding quality. Mongolian process stops offering brides. The ultimate result of this market mechanism change was to increase the production by retaining the older goats which should have gone for culling and at the expense of the quality of the fibre.

"The government intervention in the agricultural sector in many developing countries have increased market imperfection and undermine the need for processors to develop effective strategic links with the supplier".

Market pricing imperfection that affect direct trade and pricing of cashmere are of particular importance, as they have single most direct impacts on the income of herders and overall economic growth and competitiveness of the country's economy.

Many countries have introduced export ban on raw material thinking that the increasing processing capacity of the local industry will create manufacturing jobs and foreign exchange as export rises. When export are restricted the domestic price of a commodity falls affecting income of raw materials supply. On the demand side this encourages inefficient processing industries, that maybe value reducing.

To protect the domestic industry from external competition and to increase the domestic value addition the government introduced in 1994 an export ban on raw cashmere and washed cashmere. This banned encouraged many foreign processors to locate processing facilities in Mongolia and resulted in the establishment of many dehairing plants.

Between 1990 and 1993, the raw cashmere production was liberalised. Export of raw cashmere was controlled by the government.

The controlling factors are as follows:

Export licensing which is must and other trade restrictions. In 1994 export ban on raw cashmere was introduced and later in 1996 this have been replaced by export tax.

Overall the effects of export ban between 1994 and 1996 added with the tax from 1997 to present have not made the objective of the government which was to encourage better processing and exports of raw cashmere.

The tax have affected farm income, overall export earnings, the government revenue and the quality of cashmere.

The government decided to assist in improving rural herder's market excess establishing a whole network system.

Government facilitated wholesale network marketing which is another important step towards development of pashmina marketing in Afghanistan, China and Mongolia.

13.2   China

The cashmere industry has rapidly expanded in post reform era. Cashmere can be exceptional in that; China now dominate all level of value chain and provides an example for how the value chains are reformed in China and the policy implications which may arise.

The increase in cashmere production in China, the shift of processing sector to China and with mask marketing, cashmere has become a much more affordable and accessible consumer item.

All the European traders and Mongolian processors are claiming that China wants to control and regulate entire GVC for cashmere, cashmere fibres international.

Reference – de Weijer, 2007, Cashmere Fibre International.

It appears that the Chinese pashmina industry has not integrated more closely and formally with global cashmere industry despite their global market share. Chinese association and companies are not even the members of some international Cashmere agencies. It would be beneficial to Chinese agencies to understand more on standards and testing region if the formal relationship between international and Chinese industry agencies is established. Within China, Chinese policy makers have taken typically proactive and interventionist industry development efforts.

In 2009, the former prime minister of China Wen Jiabao created a working party of eight members from eight different government departments to study the problems and prospects of the industry to review the national and local government policies measures to stimulate the production of finer cashmere large scale processing (industrial parks and preferential policies in Ningxia and Inner Mongolia) and policies targeted at their production and processing sectors.

There would also seem to be more tightly specified quality characteristics in standards. Grading system has been created to cashmere purchases at farm level is a major challenge because of the large number of collector's involved and vast geographical scope.

Cashmere is the major source of cash income for herders in Mongolia and Afghanistan for many of the seven million pastoralists in western China that belongs to ethnic minorities and have an incidence of poverty that ranges from 41% in Xinjing to 20% in Inner Mongolia. Till 2010, normal cashmere prices increased by annual average of 10% in China and 11% in Mongolia. Chinese cashmere enjoyed a price over Mongolian cashmere of around 20% for a period due mainly to the quality characteristics of the Cashmere.

The cashmere industry, both internationally and in China have a highly concerned about this trend (China Animal Husbandry Product Marketing Association and China Husbandry Product Association, personal communication) (Cashmere And Camel Hair Manufacturers' Institute). China has decided to adjust to this trend through several interventions.

13.3   India

The central wool development board, Jodhpur, India has two short training projects each in favour of Ladakh autonomous Hill Development Council (HDC) and Kargil district to add value to pashmina product. A release in support of this program reports that under yield 40 trainees are trained in 4 batches for 3 months, each duration in manufacturing of wooden hand loom projects which includes the pashmina industry also. There are 245 thousand numbers of goats which are producing pashmina fibre in India. Some of the origin of the goats are Changra and Chego goats are the major producers of pashmina fibre in India.

The scheme for pashmina wool development is designed to make a meaningful intervention given the potential of this area to produce pashmina of finer quality. During the 12th plan a special package with adequate financial commitments from the government side to do the following for the pashmina development plan are:

1. To provide the necessary inputs for the breed improvement training, health care and nutritional supplement for the pashmina goats.

2. To increase pashmina population at least to a level of double the number we have now.

3. Establishment of multipurpose extension centres for nomads on migratory groups.

4. Undertake research and development work for pashmina fibre.

5. To strengthen the income from pashmina wool of the goat's breeders and sustain their interest in this activity.

6. To strengthen the existing breeding farms and develop fodder banks.

7. To strengthen the Pashmina dehairing plants at Leh.

This project will probably be handled by the Ladakh autonomous hill development council with the help of the state government.

The following are the organisation which are attached to Pashmina Wool Development Program:

1. Constitution of livestock purchase committee – Livestock purchase committee will visit villages by first week of July for screening and physical tagging of potential animals.

2. The district collector will monitor the implementation of each project meticulously and monthly report to the ministry by 5th of every month. The committee for monitoring implementation and progress of the pashmina project will be of the following members – chairmen LAHDC, deputy commissioner Leh, execute director – central wool development board, deputy secretary/director (wool) as a member, district husbandry officer (Leh).

3. The execute committee consisting of the following will have the rights to approve the project finally:

    (a) Joint secretary (wool) – Ministry of textile as chairmen,

    (b) Deputy Secretary (finance) – mot,

    (c) Director wool research association,

    (d) Director central sheep and wool research institute,

    (e) Director – animal husbandry,

    (f) Secretary General of Indian woollen mills federation,

    (g) Director planning commission of India,

    (h) Executive director central wool development board.

There are several other organisations which are involved in this process because we need to investigate all aspects of the pashmina goat to fibre herding and processing.

The organisation mentioned are only examples of how the constitutions are formed.

13.4   Afghanistan

The pro-poor strategies of Afghan government include:

1. Support the bargaining position of the producers to obtain good prices for the pashmina.

2. Improve efficiency of the supply chain to obtain a better price.

3. Promote value adding activities at the farmer's level.

4. Encourage the development of high trust maker relationship between all the actors in the markets.

5. Promote business practices that meet environmental, labour and social standards.

Key strategies to achieve these strategies:

1. Increase the volume of Afghan cashmere that reaches the market.

2. Improve the quality of Afghan cashmere.

3. Increase the efficiency of cashmere assemble and processing system with optimal benefits acquiring to the pole.

4. Improve Afghanistan marketing position and create direct end markers and linkages.

References

1. Sheikh (2016), International Journal of Advance Research, DOI: 10.21474/IJAR01.

2. Numaan (2016), International Journal of Advance Research, DOI: 10.21474/IJAR01.

3. Sibtaib (2016), Research Scholar, The Business School, University of Kashmir.

4. De Weijer, F. (2005), National Multisectoral Assessment on Kuchi.

5. Interim Afghanistan National Development Strategy (i-ANDS), 2016: www.ands.gov.af.

6. Kelley, N., "Is Pashmina pulling the wool over our Eyes?" Magazine, Page number 41.

7. LeCraw, et al. (2010), 'A Value chain Analysis of the Mongolian cashmere industry', Mongolian Cashmere Economy and Economic policy reform and competitiveness project, Chemonics Int.

8. Presentation Given at the regional cashmere conference in Bishkek, 2007.

   - J. Dugree, Economic Policy Reform and competitiveness project.

   - Dr. Ho Phan, German Wool Research Institute DWI, University of Aachen, Germany.

   - Dr. K. Spilhaus, President of Cashmere and Camel Hair Manufacturers Institute, Boston, USA.

   - B. Baatar, CEO of Goyo Fine Cashmere and Camel Hair L.L.C, Ulaanbataar, Mongolia.

9. Mishra, C.P. Allen, T. Carthy, M.D. Madhusudan, A. Bayarjagal and H.H.T. prins, 2003.

10. Mishra C., S. Bagchi, T. Namgil and Y.V. Bharatnagar, 2010.

11. Production and socio-economic changes in Indian Changhthan: Implication for natural resource management. Natural Resources Forum 34.222.230.

12. Olsan, K.A. et al. (2010), Death by a thousand huts? Effects of house-hold presence on density and distribution of Mongolian gazelles. Conservation Letters 4:1–9.

13. Spaeth, D.F. et al. (2001). Incisor arcades of Alaskan moose: Is dimorphism related to sexual segregation? Alces 37:217–226.

14. U.S. Department of Agriculture (USDA) (2008), Market Trade data. Reporting countries impact statics. 1994–2008.

15. Li XJ, Miao AD. The use of digital image processing technology in fiber detection. Testing and Standards, 2007: 62–64 Chinese.

16. Wang L., Avtar Singh and Wang X, Fibres Polym.

17. Ishrat, Y., Sofi, A.H., Wani, S.A., Seikh, F.D., and Nazir, A.B. (2011) Indian J Traditional Knowledge 17:203.

18. Raja, A.S.M., Shakyawar, D.B., and Sofi, A.H., Trends in Ruminant production.

19. Lal Chagan, Raja, A.S.M., Prateek, P.K., B Sharma, K.K., and Sharma, M.C.

20. Lal Chagan, Sharma, M.C., and P K Achives.

21. Zhang, J., Stuart Palmer and Xungai Wang, Identification of animal fiber with wavelet texture analysis, Proceedings on the world congress on Engineering (Newsletter).

22. Himlyn, F.P., Identification of animal hair fibres using DNA profiling techniques.

23. Dr. S K Chaudhuri, Regional Director South-East Asia, The Woolmark Company.

24. Pathak and Anamika (2004), Pashmina Roli Books.

25. Hunter, L. (1993), Mohair: A review of its properties, processing and applications.

26. Caulfeid, S.F.A., Saward, B.C. (1885), The Dictionary of Needle work.

27. Subramanium et al., 2005; Mccarthy, 1990

28. Lemon, Jane (2004). Metal Thread Embroidery.

29. Quinault, Marie-Jo (2003) Filet Lace Introduction to the linen stitch.

30. Van Niekerk, Di – A perfect World in Ribbon Embroidery and Stumpwork, ISBN 0-89577-059-8.

31. Wilson, David M. (1985). The Bayeux Tapestry. Thames and Hudson. ISBN 0-500-25122-3.

32. ISO 17751-1 (2016) Textile Quantitative Analysis of Cashmere, Wool, Other speciality animal fibers and their blends.

33. Coelho, L.S., Modesto, E.C, Carvalho, F.F.R., and de souse, S.I.G. (2018), Characterization and identification of Cashmere in Goats in northeastern Brazil.

34. Izuchi, Y., Tokuhara, M., (2013) Peptide profiling.

35. Ji, W., Bai, L., Ji, M. and Yang, X.A. (2011).

36. Kumar, R., Shakyawar, D.B., Pareek, P.K., Raja, A.S., Kumar, S., and Naqvi, S.M. (2015).

37. Tang, M., Zhang, W., Zhou, H., Few, Yang, J., Lu, W., Zhang, Time PCR method for quantifying mixed cashmere.

38. Gill, Slyadnev, M.N and Strongnov (2016) Optimisation of paramagnetic based nucleic acid isolation for high throughput potato pathogen detection using microchip real time PCR.

39. Nikitin, M.M, Statsyuk, N.V., Frantsuzov, P.A. (2018) Matrix approach to the simultaneous detection of multiple potato pathogens by real time PCR.

Dr. Sailen Kumar Chaudhuri obtained his M.Tech Degree in textile technology from IIT Delhi with distinction. He did his Ph.D from University of Manchester, UK. His subject of specialisation is wool and animal fibres. He is currently the Vice-Chairman (Global) and fellow of the textile institute Manchester, along with a parallel responsibility of the Indian National Office, Textile Institute, Manchester. He was the former Regional director, India and south-east Asia in the Woolmark company for 26 years. His contribution in the growth and development of Indian wool textile industry is immense. He is awarded by The Indian Woollen Mill's Federation in 1999 for his outstanding services for development of wool industry in India. For his outstanding contribution for the national development, he is awarded by IIT Delhi Alumni association in 2014. In 2014 same year, he also received an international award "Institute Medal" from the Textile Institute, Manchester, for his contribution in International development of textile industry and trade in general.

He is also an expert on Graphology, which is a science of Handwriting Analysis.